PHYSICS PRACTICAL
FOR ENGINEERS
WITH VIVA-VOCE

PHYSICS PRACTICAL FOR ENGINEERS WITH VIVA-VOCE
15 CLASSIC PHYSICS LAB EXPERIMENTS FOR ENGINEERING STUDENTS

Chandra Mohan Singh Negi

Nanhi Pari Seemant Engineering Institute,
Pithoragarh India

BrownWalker Press
Irvine & Boca Raton

Physics Practical for Engineers with Viva-Voce:
15 Classic Physics Lab Experiments for Engineering Students

BrownWalker Press / Universal Publishers, Inc.
Irvine, California & Boca Raton, Florida • USA
www.BrownWalkerPress.com
2018

ISBN: 978-1-62734-701-3 (pbk.)
ISBN: 978-1-62734-702-0 (ebk.)

Cover design by Ivan Popov.

First published in 2017 by
Overseas Press India Private Limited, New Delhi, India

Publisher's Cataloging-in-Publication Data

Names: Negi, Chandra Mohan Singh, author.
Title: Physics practical for engineers with viva-voce : 15 classic physics lab experiments for engineering students / Chandra Mohan Singh Negi.
Description: Irvine, CA : BrownWalker, 2018. | Previously released in 2017 by Overseas Press, New Delhi, India.
Identifiers: LCCN 2018940984 | ISBN 978-1-62734-701-3 (pbk.) | ISBN 978-1-62734-702-0 (ebook)
Subjects: LCSH: Physics--Laboratory manuals. | Physics--Textbooks. | College textbooks. | BISAC: SCIENCE / Physics / General. | SCIENCE / Laboratory Techniques.
Classification: LCC QC35 .N44 2018 (print) | LCC QC35 (ebook) | DDC 530.078--dc23.

**Dedicated
to
my Parents**

List of Experiments

1. To determine the wavelength of monochromatic light by Newton's ring.
2. To determine the wavelength of monochromatic light with the help of Fresnel's biprism.
3. To determine the focal length of two lenses by nodal slide and locate the position of cardinal points.
4. To determine the specific rotation of cane sugar solution using polarimeter.
5. To determine the wavelength of spectral lines using plane transmission grating.
6. To determine the specific resistance of a given wire using Carey Foster's bridge.
7. To study the variation of magnetic field along the axis of current carrying- circular coil and then to estimate the radius of the coil.
8. To verify Stefan's Law by electrical method.
9. To calibrate the given ammeter and voltmeter by potentiometer.
10. To study the Hall effect and determine Hall coeffcient, carrier density and- mobility of a given semiconductor using Hall-effect set up.
11. To determine the energy band gap of a given semiconductor material.
12. To determine E.C.E. of copper using Tangent or Helmholtz galvanometer
13. To draw hysteresis curve of a given sample of ferromagnetic material and from this to determine magnetic susceptibility and permeability of the given specimen.
14. To determine the ballistic constant of a ballistic galvanometer.
15. To determine the viscosity of a liquid.

PREFACE

This is one of enumerable self-help or how to books with an emphasis on Engineering Physics Practical. The basic premise of the book is that there are certain simple experiments, involving no more than rudimentary Physics laws and the very basic laws of Engineering Physics for undergraduate college engineering students. But these practical are often not done or taken lightly, for several reasons. First, people don't realize how easy they are to do. Second, and more fundamental, they are not done because it does not occur to people to do them. Finally, and tragically, no one in their elementary, middle, or high school educational experience has stressed the importance of doing them, and of course neither did they teach to do them. This book is to reveal to you what the experiments are, make them readily understandable, and by means of a very easy-to-use illustrations.

The main thing you should expect from this book is the theories and practical related small information more precisely about experiments. You will get a rudimentary understanding of the basic concepts behind the Engineering Physics experiment that governs the fundamental daily life questions that challenge us in life.

The book is divided into seven major categories and Fifteen chapters. In this book the students will find solutions to experimental obstacles normally faced by undergraduate college engineering students.

So in summary, you don't need any special background or ability to profit from this book.

Many people have come my way while I was working on the book. I would like to thank my parents for reminding me periodically that my opinion is, after all, just my opinion.

I am grateful to these people for inspirations.

- Dr. P. K. Garg, Vice Chancellor, Uttarakhand Technical University, Dehradun Uttarakhand.

- Dr. B. K. Singh, Director, Nanhi Pari Seemant Engineering Institute, Pithoragarh, Uttarakhand.

- Dr. Rajendra Dhobal, Director General, U-COST Uttarakhand.

- Dr. A. K. Swami, Ex. Principal G.B.P. Engg. College, Pauri, Uttarakhand.

- Professor P. S. Jagwan, Principal Govt. (P.G.) Colllege, Augustyamuni, Uttarakhand.

- Dr. Ankur Upadhyay, Assistant Professor, Department of Applied Sciences, Nanhi Pari Seemant Engineering Institute, Pithoragarh, Uttarakhand.

I am especially thankful to the following people for their academic and subjective support in this accomplishment.

- Dr. V. K. Jain, Vice Chancellor, Doon University Dehradun, Uttarakhand, (Ex Vice Chancellor, UTU, Dehradun).

- Dr. J. S. Gariya, Asst. Prof. LSM Govt. (P.G.) Colllege, Pithoragarh Uttarakhand.

My very sincere thank goes to Uma Bansal who has been kind enough to make the book for the students. I must acknowledge that without her help, I could not have completed the manuscript. She helped me in preparing tables, lists, and helped in checking a number of times. She being a student of Physics, she was critical and helpful while I was working on the book.

My thanks to Mr Sunil Mehta and Mr Mukesh Pandey also who did the page-making at the initial stages for the book. Finally I am indebted to Mr Surinder Lijhara, Editoral Director at Overseas Press for taking up the publication of this work.

Chandra Mohan Singh Negi

CONTENTS

OPTICS

ELECTRICITY AND MAGNETISM

SEMI CONDUCTOR PHYSICS

CHEMICAL EFFECT OF CURRENT

MAGNETIC PROPERTIES OF MATTER

BALLISTIC GALVANOMETER

PROPERTIES OF MATTER

APPENDICES

Physical Constants and Their Standard Values in Tabular Form

How to Record the Experiment

Students are advised to record the experiment on your practical note book according to the following scheme:

Day and date.......

Experiment No. ....

Object : Here the object of the given experiment will be written.

Apparatus required : The list of the instruments required for performing the actual experiment will be given here.

Diagram : The circuit diagram in case of the experiment of electricity and electronics, and any diagram in case of experiments on light is given on the left plane paper of the note book.

Formula Used :Here the formula, which is to be used for the calculation will be written along with the explanation of the symbol given in it.

Observations : Here the least count of the instrument used for the measurement along with the table for recording the experimental observations must be given.

Graph : The graph between the two independent variables must be drawn if required, on a graph paper and paste it on left page.

Calculations : Calculations must be done by using log and antilog tables and if necessary by using natural sin, cos and natural tangent tables.

Result : Here result must be mentioned with proper unit.

Standard Result : Here standard result must be written (if any)

Percentage Error : Here percentage error must be given.

Precautions and sources of errors : Here precautions taken during the performance of experiment and sources of errors are given.

The above arrangement is extremely useful in maintaining a proper record. It should be remember that the experiment affords the opportunity to learn the habit of systematic observations, to do the things honestly, efficiently and regularly. Intelligent and sincere work is more important than accurate results.

Experiment No. 1

Object : To determine the wavelength of Sodium light by Newton's Ring experiment.

Apparatus required: Newton's rings apparatus of an travelling microscope & lens, sodium bulb, sodium vapour lamp transformer & wooden stand etc.

Description of Apparatus: Newton's ring can easily be obtained in the laboratory by using the apparatus whose optical arrangement is shown in Fig. 1.1. The actual apparatus consists of a seasoned wood stand for supporting the travelling microscope M. Exactly below the objective of the microscope, an optically true plane surfaced clean glass plate is placed at the bottom of the opened wooden box. Over this glass plate a plano-convex lens L is placed in such a way that the central point of its curved surface touches the upper surface of the glass plate G. Slightly above this combination is an

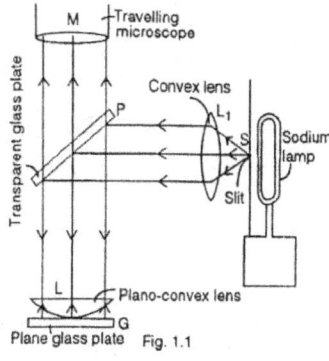

arrangement for mounting another thin plate of glass inclined at 45^0 to the vertical.

Formation of Newton's Ring : Formation of Newton's ring is shown in Fig. 1.2,

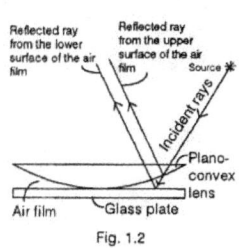

Fig. 1.2

which is a result of interference between the light rays reflected from the upper and lower surfaces of the air film between a plane glass plate and a convex side of a plano-convex lens in contact with it (Fig. 1.2). Light from an extended monochromatic source S rendered parallel by a lens L_1 and then falls on a 45^0 inclined thin transparent glass plate P which partially reflects the light in the vertically downward direction. These reflected beams fall normally on an air film formed between the convex surface of the plano-convex lens and glass plate G. The light transmitted through the plano-convex lens on reflection from the surface of the air film in contact with the glass plate interferes with the light reflected from the surface of the air film in contact with the lower surface of the plano-convex lens. These reflected beams proceed upward and enter the observer's eye through a low travelling microscope M. On focusing, a large number of Newton's rings, alternately bright and dark can be seen in the field of view of the microscope.

Formula used : If D_{n+p} and D_n are the diameters of $(n+p)^{th}$ and n^{th} dark (or bright) rings respectively, then the wavelength of monochromatic light is given by

$$\lambda = \frac{D_n^2}{\underline{\hspace{3cm}}}$$

where R = Radius of curvature of plano-convex lens

and   p= Difference of order of ring under consideration.

Procedure :

1- First of all the glass plate P and G, and plano-convex lens L are thoroughly cleaned with spirit.

2- The plano convex lens is placed on the plane glass plate G such that its convex surface touches the glass plate. Arrange the other glass plate P at 45^0 to the vertical in the mounting.

3- Place the arrangement in front of a sodium lamp so that the height of the centre of the glass plate P is the same as that of the centre of the sodium lamp. Place a screen in between having a hole of about one inch square in it at same height. Place the convex lens in between & adjust its position so that a parallel beam of light in made to fall on the glass plate P at angle 45^0.

4- Adjust the position of the microscope so that it lies vertically above the centre of the lens focus the microscope, so that alternate dark and bright fringes are clearly visible.

5- Slide the microscope to the left and right till the crosswire lies at the centre of 20th dark ring. Such as shown in Fig. 1.3.

6- Note the reading on the vernier scale of the microscope when the cross-wire lies tangentially at the centre of the 16th, 12th, 8th and 4th dark rings respectively in both sides.

Fig 1.3

7- Remove the lens L and find the radius of curvature of the surface of the lens in contact with the glass plate G accrately using a spherometer.

Observation Tables:

(1) Table A : Determination of Diameter

Least count of the microscope $= \dfrac{\text{f one division of mair scale}}{\text{vision on the vernier scale}}$

No of rings	S. No	Left end of the Ring L			Right end of the Ring R			L-R (D_n)	$(L-R)^2$ $D_n^{\ 2}$	D^2_{n+p} $-D^2_n$
		Main scale Reading (a)	Vernier Reading (b)	a+b	Main scale Reading (a')	Vernier Reading (b')	a' +b'			
	1									
	2									
	3									
	4									

(2) Table B : Measurement of the Radius of Curvature of Plano-Convex Lens

Least count of the spherometer $= \dfrac{}{\text{of divisions on the circular scale}}$

$$= \frac{\text{.....}}{\text{....}} = \text{.....cm}$$

Pitch $= \dfrac{\text{e moved along circular scale}}{\text{r of complete rotations given to the circular scale}}$

$$= \frac{\text{....}}{\text{....}} = \text{....cm}$$

S.No.	Initial circular scale Reading on the plane glass plate (a)	No of complete rotations given to the circular scale (n)	Final circular scale reading on the convex surface of the lens (b)	Number of additional circular scale division moved [m=(b~a)]	h=(n X pitch+ m X least count) (cm)
1.	...	...	...	...	...
2.	...	...	...	...	...
3.	...	...	...	...	...
4.	...	...	...	...	...
5.	...	...	...	...	...
				Mean h	.......cm

Usually the radius of curvature (R) of the convex surface of the plano-convex lens is given. If it is not given, then find it by the spherometer.

From the Fig.1.4,

Distance between the legs of the spherometer (l)

AB=.....cm

BC=....cm

CA=....cm

Mean l $= \dfrac{+C4}{}$ = ...cm

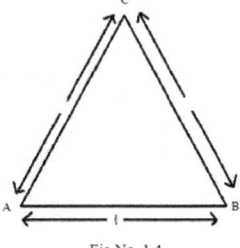

Fig No. 1.4

Calculations:

1. The radius of curvature R of the convex surface of the plano-convex lens is calculated by substituting the measured values of l and h from the observation table in the following formula,

$$R = \frac{}{} + \frac{}{} \quad \ldots\ldots\ldots cm$$

2. The wavelength of sodium light λ is calculated by substituting the calculated value of R and measured values of $D^2_{n+p} - D^2_n$ and p in the following formula,

$$\lambda = \frac{-D_i}{} = \ldots.. cm = \ldots\ldots A^0$$

Result: The wavelength of sodium light = $\ldots\ldots\ldots A^0$

Standard Result: Standard value of wavelength of sodium light = $5893 A^0$

Percentage Error: The percentage error in the experimental value is calculated by the following formula

$$\text{Percentage error} = \frac{\text{'d value} - \text{calculated value}}{\text{'d value}} \times 100\% = \ldots..\%$$

Precautions and Sources of Error:

1. The plano- convex lens and glass plate should be properly cleaned to make them dust free.
2. For the measurement of the diameter of the rings use bright rings not dark ones because the cross-wire can be eaisly be set in the middle of the bright ring rather than dark ring.
3. To find the diameter of the ring, cross-wire should be arranged tangentially on the concerning ring.
4. The micrometer should move in one direction only otherwise back-lash will appear.
5. To enable the eye to see the entire film simultaneously an extended source of light is used instead of a point source.

Viva-Voce

Q.1. What are you doing?

Ans. Sir/Madam, I am determining the wavelength of sodium light by using Newton's rings.

Q.2. What are Newton's ring?

Ans. The formation of Newton's rings is a special case of interference in a wedge shaped air film. When a monochromatic light falls normally on a wedge shaped air film developed between the lower (convex) surface of the plano-convex lens and upper surface of the plane glass plate (when placed in contact), we get an inner dark spot surrounded by alternate bright and dark rings. These rings are called Newton's rings, after the name of its inventor.

Q.3. How are these rings formed?

Ans. According to Young, Newton's rings are produced as a result of interference between the light waves reflected from the upper and lower surfaces of the air film developed between the convex surface of the plano-convex lens and the plane glass plate.

Q.4. Why are these rings circular?

Ans.: These rings are circular because the air film formed between the plano-convex lens and plane glass plate is wedge shaped and loci of the points of equal thickness are circles concentric with the point of contact.

Q.5. Why is the 45^0 plate employed?

Ans. The transparent glass plate at 45^0 turns the light rays coming from an extended source to 90^0, so that they fall normally on the plano-convex lens placed on the horizontal glass plate.

Q.6. Why is the centre of the rings dark?

Ans. At the point of contact (at the centre) of the plano convex lens and the glass plate the thickness of the air film, that is, the actual path difference between the two interfering beams is zero, but the effective path difference is $\lambda/2$ or a phase change of π. It is because one of the beams is reflected from the glass plate which is a denser medium. Thus at the centre the condition of minium intensity is satisfied. Hence the central spot of the rings is dark.

Q.7. What are the uses of Newton's rings?

Ans. Newton's rings are used for the determination of the wavelength of monochromatic light, refractive index of a liquid and the radius of the spherical surface.

Q.8. If the fringes are not exactly circular what do you infer?

Ans: If the fringes are not circular it means that either the upper surface of the glass plate or the convex surface of the plano-convex lens are irregular. Some times it is due to the presence of dust particles between the two surfaces.

Q.9. How will you get bright spot at the centre due to reflected light?

Ans. To obtain the bright centre of the rings system, we have to take a crown glass plano-convex lens and flint glass plate with a few drops of sassafras oil in between them.

Q.10. What are the two nearby wavelengths produced by the sodium light?

Ans. 5890 A^0 and 5896A^0.

Q.11. Do you get rings in the transmitted system?

Ans. Yes, in the transmitted system, we shall have a dark fringes at the place where bright fringes have been obtained in the reflected system because in the transmitted system te conditions for maximum and minimum intensities are opposite to those obtained with reflected light. Hence, the two system of fringes are complimentry. The central spot will be bright when viewed by transmitted light.

Q.12. What will happen if the film is illuminated by white light instead of yellow (sodium) light?

Ans. A series of concentric coloured rings with a dark centre will be observed around the point of contact.

Q.13. What will happen if we use a plano-convex lens of small radius of curvature?

Ans. Since the diameter of bright or dark ring is directly proportional to the square root of the radius of curvature of the lens. Therefore if we use a les of small radius of curvature, the diameter of ring will be small.

Q.14. What will happen if we use a mirror in place of a glass plate?

Ans. If we use a mirror in place of a glass plate, the transmitted system of fringes will also be reflected by the mirror and due to the superposition of the reflected and transmitted light uniform illumination will observe in the field of view of the microscope, because the conditions of maxima and minima are opposite to each other in the reflected and transmitted beams.

Q.15. What happen if the convex surface in Newton's rings apparatus is replaced by an ordinary glass plate?

Ans. In this case interference takes place but the shape of the fringes is irregular.

Q.16. How do you define the pitch and the least count of a spherometer?

Ans. The distance between any two consecutive threads measured parallel to the axis of the spherometer is known as the pitch of spherometer The distance moved along the vertical scale when we move the circular scale through one division is called the least count of the spherometer.

Q.17. What is microscope?

Ans. Microscope is an optical instrument used to magnify the image of a very small object near to it.

Q.18. What do you mean by magnifying power of a microscpe?

Ans. Magnifying power of a microscope is the ratio of the angle substended by the final image at the eye when seen through the microsope to the angle substended by the object at the eye when seen directly by naked eye, when both object and image laying at the least distance of distinct vision.

Q.19. What is the magnification?

Ans: Magnification is defined as the ratio of the linear size of the image to the linear size of the object.

Experiment No. 2

Object: To determine the wavelength of monochromatic light (sodium light) with the help of Fresnel's bi-prism.

Apparatus required: Optical bench with four uprights, sodium lamp , biprism, convex lens, micrometer eyepiece, reading lens, reading lamp, spirit level and small quantity of spirit.

Description of Apparatus: The apparatus consists of graduated optical bench of heavy metal base provided with levelling screws. The bench is provided with four uprights which hold a slit of adjustable width, a biprism, a convex lens of short focal length and a micrometer eye-piece as shown in Fig. 2.1.

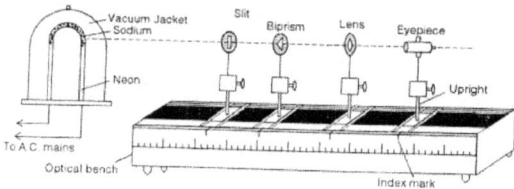

Fig. 2.1

The slit and biprism can be rotated in their own vertical planes with the help of tangent screws. All the four uprights are provided with rack and pinion arrangement and can move along the bench as well as in a direction perpendicular to the bench in the horizontal plane. A line is marked on each uprights at the base in the centre to take readings on the scale. This line is called index mark.

Formation of Virtual Sources and Fringes:

Light from a narrow vertical slit S, illuminated with monochromatic light of wavelength λ, is allowed to fall symmetrically on the vertical refracting edge of the biprism (Fig. 2.2). When this light falls on the lower half of the prism it bent upwards and appears to come from S_2.

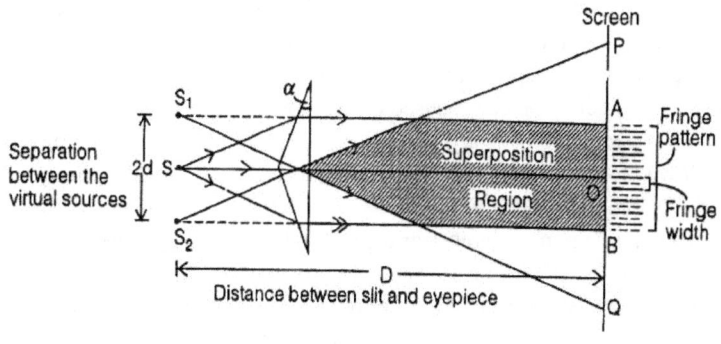

Fig. 2.2

Similarly, light from S falling on the upper half of the biprism is bent downwards and appears to come from S_1. The virtual   images S_1 and S_2 being the images of S act as two coherent sources. The source S_1 illuminates the portion AQ of the screen while the source S_2 illuminates the portion BP. In the common shaded region AB the coherent beams superimpose and, therefore, interfere. Hence, in the region AB bright and dark interference fringes are obtained.

Formula used : The wavelength λ of the monochromatic light is given by the relation

$$\lambda=$$

Where   ıge widtḫe width,

2d=distance b/w the two virtual sources, and

D = distance b/w the slit and screen Again,

$$2d= \overline{}$$

Where d_1=distance between the two images formed by the convex lens in one position, and

d_2= distance between the two images formed by the convex lens in second position.

Procedure:

1- Level the bed of optical bench with the help of spirit level.
2- The slit, bi-prism and eyepiece are adjusted at the same heights. The slit and the cross wire of eye piece are made vertical.
3- The micrometer eye piece focused on crosswire.
4- With an opening provided to the cover of the monochromatic source, the light is allowed to incident on the slit and the bench is so adjusted that light comes straight along its lengths.
5- Place the Bi – prism upright near the slit and move the eye piece sideway. See the two images of the slit through Bi – prism, if they not seen, move the upright of Bi – prism right angle to the bench till they are obtained. Make the two images parallel by rotating bi-prism in its own plane.
6- Bring the eyepiece near the Bi- prism and gave it a rotation at right angle of the bench to obtain a patch of light. As a matter of fact, the interference pattern is obtained in this patch provided that the edge of the prism is parallel to the slit.
7- To make edge of the Bi- prism parallel to the slit, the Bi- prism is rotated with the help of tangent screw till a clear interference pattern is obtained.
8- In order to adjust the system for no lateral shift, the eyepiece is moved away from bi-prism. In this case, the fringes will move to the right or left but with the help of base screw provided with bi-prism, it is moved at right angle to the bench in a direction to bring the fringes back to their original position.

Now move the eyepiece towards the Bi- prism and the same adjustment is made with the help of eyepiece. Now using the process again and again, the lateral shift is removed.

Observations:

Pitch of the micrometer screw = ---- cm

Total number of divisions on the circular scale = -----

Least count of micrometer screw = ---- cm.

1. Table A: For Fringe Width (

No. of fringes	Micrometer reading			No. of fringes	Micrometer reading			Difference for 20 frings	Fringe width
	M.S. reading (cm)	V.S. reading (cm)	Total (cm) (a)		M.S. reading (cm)	V.S. reading (cm)	Total (cm) (b)		
1.									
2.									
3.									
4.									

Mean fringe width

2. Measurement of D :

Position of the upright carrying slit = ----cm

Position of upright carrying the eyepiece = -----cm

Observed value of D = ------cm

Value of D corrected for bench error = -----cm.

3. Table B: Measurement of 2d

S.No.	Micrometer Readings						$2d=\sqrt{\text{Mean}}$	Mean
	1st Position of lens			2nd Position of lens			(cm)	2d (cm)
	Position 1st image (cm) a	Position 2nd image (cm) b	d_1 cm (b-a)	Position 1st image (cm) a'	Positio 2sd image (cm) b'	d_2 cm (b'-a')		
1.								
2.								
3.								
4.								

Calculation: The value of d is calculated by following formula

$$\lambda = \text{---- } cm \text{ c- } cm \text{ or ---- } A^0$$

Percentage Error: The percentage error in the experimental value is calculated by the following formula

$$\text{Percentage error} = \frac{d \text{ value} \sim \text{calcu} \sim \text{ed value}}{d \text{ value}} \times 100\% = \ldots \ldots \%$$

Result: The wavelength of monochromatic ligh t= ---- A^0

Standard Result: Standard value of wavelength of sodium light = 5893 A^0

Precautions and Sources of Error:

1. The slit should be made as narrow as possible.
2. The slit should be parallet to the vertical cross–wire of the eye- piece.
3. Bench error should be taken into consideration while measuring D.
4. The line joining the centre of the slit and the edge of the biprism should be exactly parallel to the length of the bed of the bench.

5. A convex lens of short focal length should be used to determine 2d.
6. Cross-wire should be settled at the centre of the frings while taking observation for frings width.

Viva-Voce

Q.1. What are you doing?

Ans. Sir/Madam, I am determining the wavelength of monochromatic light with the help of bi-prism.

Q.2. What do you mean by monochromatic light?

Ans. The light which has only single wavelength is called monochromatic light. Practically sodium light is taken as monochromatic.

Q.3. What is bi-prism?

Ans. The bi-prism consists of two acute angled prism with their base in contact. It is actually a simple prism of obtuse angle of 179^0 and acute angle on both side is $(1/2)^0$. This is made from an optically plane glass plate by proper grinding and polishing.

Q.4. Why are the refracting angles of the two prism made so small?

Ans. The refracting angles of the two prisms are made very small because for small refracting angles the separation 2d between the two virtual images of the slit will be small and so the fringes width will become sufficiently larger. Large separation is necessary for accurate measurement.

Q.5. What do you mean by interference of light?

Ans. When the two light waves of same frequency and constant phase difference between them pass through the same region of medium and superimpose with each other, there is a modification in the intensity of light in the region of superposition. This modification in the intensity so obtained by the superposition of the waves is called interference.

Q.6. What are interference fringes?

Ans. The interference frings are the narrow patches of light of maximum and minimum intensities in the region of superposition of two waves.

Q.7. What are coherent sources?

Ans. The two sources of light which emit waves with a zero or constant phase difference between them are called coherent sources.

Q.8. Why should the slit be narrow?

Ans. The slit is made narrow to obtain good contrast of the interference fringes.

Q.9. What is fringes width?

Ans. The distance between any two consecutive bright frings or any two consecutive intensities in the region of superposition of two waves. For bi-prism, it is mathematically expressed as follows:

where λ is the wavelength of light, D is the distance between the slit and the screen and 2d is the distance between the two virtual sources.

Q.10. What will happened if in your experiment sodium yellow light is replaced by blue light?

Ans. If the sodium yellow light is replaced by blue light the frings width decrease and fringes become more densed.

Q.11. In your experiment where are the two coherent sources situated?

Ans. In this bi-prism experiment two coherent sources are situated symmetrically on either side of the silt and in the same plane as slit.

Q.12. What effect do you observe on the screen due to interference of light?

Ans. A region of alternate high and low intennsities are seen. The respective regions of alternate high and low intensities are called bright and dark fringes and region as a whole is called interference pattern.

Q.13. Is there any loss of energy in interference?

Ans. No, there is no loss of energy but energy is simply redistributed.

Q.14. Are the bi-prism fringes perfectly straight?

Ans. The bi-prism fringes are not perfectly straight but hyperbolic. The eccentricity of the hyperbola is so large that they appear straight alternate bright and dark in the field of view of the eye piece.

Q.15. What will happen if you use a white light source instead of sodium light?

Ans: If we use a white light source, an interference pattern of coloured frings with a white central fringe will be seen in the field of view.

Q.16. Can you locate the zero order fringe?

Ans. Yes, first of all interfference fringes are obtained with sodium light and then by replacing the sodium light with white light we can locate zero order fringe. In the interference pattern with white light zero order fringe appear white, while others are appeared coloured.

Q.17. Can you locate the zero order fringe with yellow sodium light?

Ans. No, because with sodium light all the fringes appear same without any distinction.

Experiment No. 3

Object: To determine the focal length of the combination of two lenses separated by a distance with the help of a Nodal slide and then to verify the formula

$$\frac{1}{F} = \frac{1}{f} + \frac{1}{f_1} - \frac{x}{f\,f}$$

Where F=Focal length of the combination of two convex lenses of focal lengths f_1 and f_2

and   x=distance between two lenses

Apparatus required: Nodal slide assembly consisting of an optical bench comprising with four uprights- a bulb in a metallic container, cross-slits screen, nodal slide and plane mirror, and two convex lenses of nearly same and of short focal lengths.

Description of Apparatus: The apparatus consists of one and a half meter long optical bench of heavy metal base comprising with four uprights as shown in Fig.3.1. On one upright, fixed at one extreme end of the bench, a bulb in a metallic container

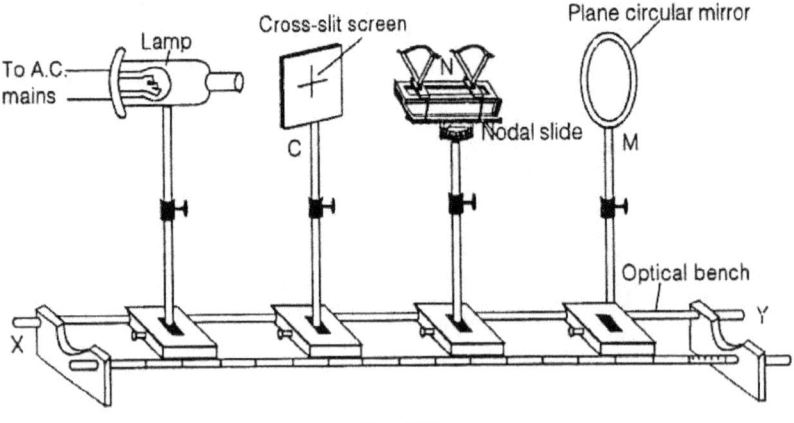

Fig. 3.1

which has a circular opening for the emission of light is mounted. The light rays emitted by the bulb illuminate the cross-silts made on a metal plate screen mounted on the adjacent upright. Next to the cross-slits upright is an another another upright on which the nodal slide is mounted. As evident from the Fig. 3.2, the nodal side is a small horizontal metal carriage having two holders in which the two lenses are placed. The seperation between the lens holders can be changed by moving them axially and their positions can be directly read on the linear scale provide with the carriage. The metal carriage as a whole can be moved

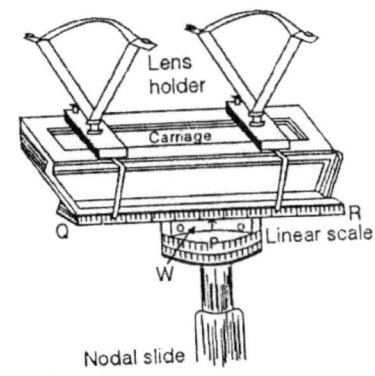

Fig No. 3.2

back and forth in horizontal direction with the help of a rack and pinion arrangement so that the relative positions of the two lenses can vary relative to the axis of the upright. The carriage is also capable of rotation about a vertical axis and its angle of rotation can be recorded from its graduated circular base. The fourth upright carries a circular plane mirror which is capable of rotation about a horizontal axis perpendicular to the bed of the bench and can be tilted in its own plane. All these four uprights are arranged on the bench in such a way that on one side of the nodal slide, there is a lamp and a cross-slits, and on the other side there is a plane mirror as shown in Fig. 3.1.

Principle and Theory: The function of the nodal slide is based on one property of nodal points of an optical system. According to this property, if an incident ray passes through first nodal point N_1 of an optical system, then after refraction through the system, the refracted ray necessarily emerges through its second nodal point N_2 in a direction parallel to the original direction.

On the basis of this property, the principle of nodal slide may be stated as follows:

If a beam of parallel rays is incident on any optical system of converging lens, an image of an object is formed on a screen placed at second focal plane [Fig.3.3 (a)].

If the system is slightly rotated about a vertical axis passing through its second nodal point (N_2) the image does not shift laterally [Fig.3.3 (b)]. Conversely, a lateral shift

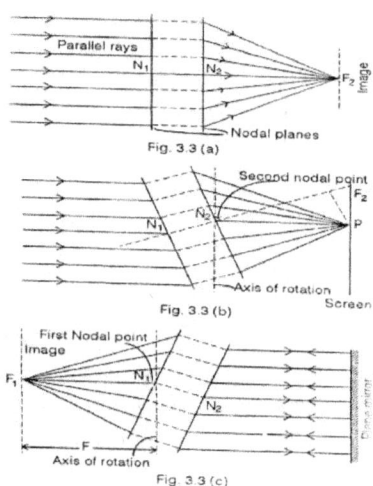

Fig. 3.3 (a)

Fig. 3.3 (b)

Fig. 3.3 (c)

in the image will be observed if the optical system is rotated about any point other than nodal point.

The similar phenomenon will occur if the optical system is slightly rotated about a vertical axis passing through its first nodal point (N_1), that is, the image formed on a screen placed at first focal point F_1 or at a first focal plane remain unaffected by the rotation of the system about the vertical axis passing through its first nodal point [Fig.3.3 (c)]. In this situation, if a plane mirror is placed normally to the emergent rays, then the rays are reflected back to the same path and the image of the object placed at F_1 is formed on F_1 itself [Fig.3.3 (c)]. In this way, the positions of the nodal points and corresponding focal points can be located with the help of a nodal slide.

As the medium on either side of the lens system is same (or air), the principal points coincide with the corresponding nodal points. The focal length of such an optical system of converging lenses becomes the distance between the nodal point and the corresponding focal point. Hence, the distance of the screen situated at F_1 (or F_2) from the axis of rotation passing through the nodal point N_1 (or N_2) gives the principal focal length of an optical system.

Formula used: If f_1 and f_2 are the focal lengths of the two given convex lenses, separated by a distance 'x', then the focal length of the combined system is given by the relation

$$\frac{1}{F} = \frac{1}{f} + \frac{1}{f_1} - \frac{x}{f\ f}$$

where

F=Focal length of the combination,

f_1, f_2= Focal length of given lenses

and x=Seperation between two lenses.

Procedure:

1. Mount lamp, cross-slits screen, nodal side and the plane circular mirror on the their respective uprights and adjust them to the same height.
2. Switch on the lamp and adjust the lens on nodal slide. The light passing through the cross- slit is made parallel by the lens and allow to fall normally on the plane mirror, which after reflection return on its own path and focus on the cross-slit screen.
3. In this setting, note the position of cross- slit and position of lens given the first focal length.
4. Now rotate the nodal slide through 180^0, so that the other face of the lens faces the incident light, now repeat f in this situation. Finally find the mean focal length f_1.

5. Remove the first lens and mount the second lens length f_2 and repeat the above procedure for both face of the lens to of focal determine the focal length f_2 finally find the mean focal length f_2.

6. Now mount both the lenses on the nodal slide and keep them some distance (x) apart. In this setting repeat the above step (2 to 3) and find focal length F in one face and other face.

Observation Tables:

(1) Table A: For measurements of focal length of lenses (f_1,f_2)

S.No	Light incident on	For lens L_1					For lens L_2				
		Position of cross-slit upright (a) (cm)	Position of lens upright (b) (cm)	Observed focal length f_1= a-b cm	Corrected Focal Length $f_{1(cm)}$	Mean focal length $f_1(cm)$	Position of scross-slit (a')	Position of lens(b')	f_2= a'-b'	Corrected Focal Length	Mean (f_2)
1	One face										
	Other face										
2	One face										
	Other face										
3	One face										
	Other face										

(2) Table B: For focal length F of the lens combination

S.No	Light incident on	Distance between two lenses (x)	Position of cross-slit upright (a) (cm)	Position of axis of rotation of Nodal slide (b) (cm)	F=a-b Observed focal length of combination F=(a-b) (cm)	Corrected focal length of combination F (cm)	Mean (F) cm
1	One face	x_1....					
	Other face						
2	One face	x_2					
	Other face						
3	One face	x_3					
	Other face						

Calculation: From Table A

$$f_1 = \text{.... cm}$$

$$f_2 = \text{..... cm}$$

$$\frac{1}{F} = \frac{1}{2} + \frac{1}{\iota J} - \frac{x}{}$$

Or $F_1 = \text{..... cm}$

Similarly, for $x = x_2$, $F_2 = \text{..... cm}$

for $x = x_3$, $F_3 = \text{..... cm}$

Mean value of focal length F of combination of two lenses

$$\frac{1}{-} = \frac{+\frac{7}{3}}{} = \ldots cm$$

Result:

The calculated and experimentally determined values of focal length F of combined lens system is approximately equal, hence, the lens formula

$$\frac{1}{F} = \frac{1}{?} + \frac{1}{lJ} - \frac{x}{}$$

is verified.

Precautions and Sources of Error:

1. All the upright arranged on the optical bench should be adjusted to the same height
2. The mirror used should be truly plane.
3. The cross-slit must be properly and intensely illuminated.
4. The positon of no shift, in which the image does not laterally on rotating the carriage, should be precisely determined.
5. For obtaining well defined and sharp images of the cross-slit screen, the aperature of the lens or lenses should be taken small.
6. For searching nodal points on the principal axis of the lens system or lens, the rotation of the nodal slide about the vertical axis should not exceeded by 5^0 or so.
7. Bench error should be properly accounted for.

Viva-Voce

Q.1. What are you doing?

Ans. Sir/Madam, I am determining the focal length of a coaxial optical system of two thin convex lenses seperated by a distance with the help of nodal silde.

Q.2. What is nodal slide?

Ans. Nodal slide is a small horizontal metal carriage having two lens holders. The seperation between the lens placed in the lens holders can be changed and directly read on the linear scale provided with the carriage (Fig. 3.2). The metal carriage as a whole can be moved back and forth in horizontal direction as well as can be rotated about a vertical axis.

Q.3. What are nodal points?

Ans. A pair of conjugate points on the principal axis of the optical system having unit positive angular magnification are called nodal points of that optical system. It means that if a ray of light is incident on one of these points, after refraction through the optical system it emerges from the other nodal point parallel to the original direction.

Q.4 . What do you mean by focal planes?

Ans. The plane passing through the first focal point and perpendicular to the principal axis of the lens system is called first focal plane. Similarly, the plane passing through second focal point and perpendicular to the direction of the principal axis of the lens system is called second focal plane.

Q.5. What is the principle of your nodal slide?

Ans. Nodal slide is based on the principle based on the property of nodal points. If a beam of light parallel to the principal axis is incident on a coaxial converging lens system and after refraction through the system image is formed on a scren placed at the second focal plane, then the image of the object does not shift laterally when the system is slightly rotated (approximately 5^0) about a vertical axis passing through its second nodal point.

Q.6. Is it necessary to rotate the lens combination by 180^0?

Ans. Yes, beacause the nodal points are not symmetrical with respect to two lenses of different focal lengths. But if the focal length of the two lenses are same, there is no need to rotate the lens combination by 180^0.

Q.7. Who discovered the Nodal points?

Ans. The Nodal points were discovered by Listing.

Q.8. If the position of plane mirror is changed, how will the position of the final image be affected?

Ans. The position of the plane mirror does not affect the position of image, it is simply used to focus the image on the cross-slit itself.

Q.9. In your experiment, why more than one images of the cross-slit are appearing?

Ans. In addition to the actual image of the slit few false images are appearing due to partial reflections from the face of the lens.

Q.10. What is the function of plane mirror in your experiment?

Ans. The plane mirror simply reflects the parallel beam of light falling on it.

Q.11. What are cardinal points of an optical system?

Ans. There are six cardinal points of an optical system, viz. (i) two focal points, (ii) two principal points, and (iii) two nodal points.

Q.12. How will you locate the true image out of these false images?

Ans. It can be recognize by one of the following two ways: (i) By giving slight rotation to the plane mirror: If by giving slight rotation to the mirror the image moves on the screen then it is the true image otherwise the image is false. (ii) If we simply put our hand in front of the mirror, the true image will disappear and false images remain.

Q.13. How will you define nodal planes?

Ans. The planes passing through the nodal points and perpendicular to the principal axis are called nodal planes of the coaxial optical systems. Corresponding to two nodal points, there are two noal planes

Q.14. When the system is dipped in water, is the relative positions of nodal and principal points of your lens system changed?

Ans. No, because still the medium on either side of the system is same.

Experiment No. 4

Object: To determine the variation of magnetic field along the axis of a current carrying coil and then to estimate the radius of coil

Apparatus required: Stewart and Gee type tangent galvanometer, storage battery, commutator, ammeter, rheostat, connection wires, one way plug key and a piece of sand

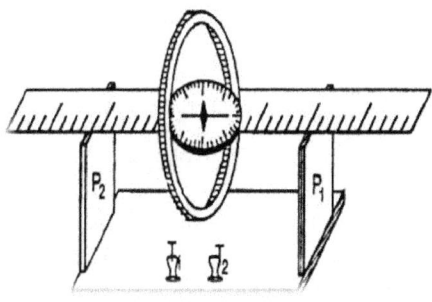

Fig. 4.1

Description of Apparatus: Stewart and Gee type tangent galvanometer consists of a large number of turns of insulated copper wire wound over the groove of a circular wooden or brass frame fixed on a horizontal base with its plane in a vertical direction as shown in Fig.4.1.

The ends of coil are connected to the two binding terminals T_1 and T_2 provided at the base of the instrument. A deflection magnetometer compass box is placed on a bench fixed on two vertical pillars P_1 and P_2 in such a way that magnetometer box can slide along the horizontal direction with centre of the magnetic needle always lies on the axis of the coil.

Two meter scales opposite to each other are firmly attached with the bench of the magnetometer for measuring the distances of the needle from the centre of the coil on either side.

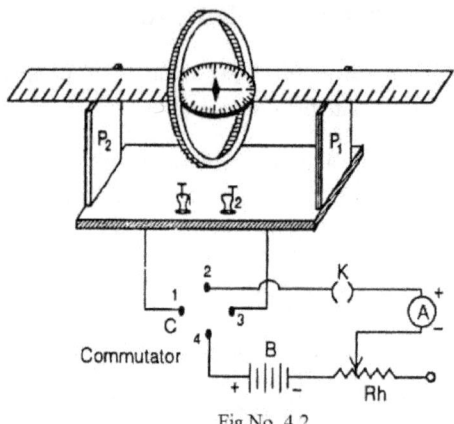

Fig No. 4.2

The scheme of connections is clearly depicted in the diagram in Fig. 4.2.. The binding terminal T_1 and T_2 of the tangent galvanometer are connected to the two diagonally opposite fixed terminals (1,3) of commutator C. The remaining two movable terminals (2,4) of commutator are connected to the storage battery B with a rheostat Rh, an ammeter A and key K in series as shown in Fig. 4.2.

Theory and Formula used: The magnetic field F at any point distant x from the centre of the coil along its axis is given by

$$F = \frac{}{+r \cdot (^{3/2}\quad) \; /}$$

Where,　I = current flowing through the circular coil,

　　　　r = radius of coil

and　　n = number of turns in the coil

But　　F = H tan

$$\therefore \quad \frac{}{+r \cdot (^{3/2}\quad) \; /}n = H \tan\theta$$

or $\overline{+r(3/2)}$ ∝ tanθ

Hence the variation of magnetic field along the axis of a current carrying circular coil involves the plotting of a graph between x and tanθ.

Procedure:

1. First of all the magnetometer compass box is placed on the bench such that its magnetic needle lies at the centre of the coil. Rotate the instrument in the horizontal plane till the plane of the coil lies roughly in the magnetic meridian or bench in the East-West direction. Now without disturbing the setting of the apparatus rotate the compass box till pointer read 0^0-0^0 on the circular scale. This adjustment of the instrument remain unchanged throughout the experiment.

2. Insert the plug in the key K so that the current flows in the coil. Adjust the value of current with the help of rheostat such that the deflection in the galvanometer lies between 70^0- 75^0. Note this value at both ends of the pointer. Reverse the direction of current in the coil with the help of the commutator and again note the reading of both the pointers.

3. Now slide the magnetometer box along the axis to get maximum deflection. In this situation the centre of the needle coincide with the centre of the coil. Note this position of magnetometer box on the meter scale.

4. Now shift the position of compass needle box along the axis in equal step of 2 cm on one side of the coil along the bench. Note this distance and also the reading of both ends of the pointer for direct as well as reversed current. This process of shifting of compass in small step of 2cm and taking the observation of ends of the pointer for direct and reversed current, is continued till the deflection is reduced to 30^0. Note all these reading in table. It should be remember that during each observation the current in the circuit always remain constant.

5. Now repeat the process of taking observation on other side of the coil keeping current always constant.

6. Plot a graph between the distance x of the compass needle box from the centre of the coil or from the one end of the bench and tangent of deflection of magnetic needle, that is, tanθ. The resulting curve shall be symmetrical and its

maximum value shall correspond to the position of the needle at the centre of the coil itself [Fig.4.3].

7. Find out the points of inflexion A and B on the curve by drawing common tangent at the place where the curve is practically a straight line for a short length and measure the distance between them to get the radius of the coil.

Observations:

(1) **Table A: For distance x from the centre of the coil and tangent of deflection of the magnetic needle, that is, tanθ**

S. No.	Distance of the needle from the centre of the bench (x) (cm)	Deflection on the left side of the coil in degree				Mean	Tan θ	Deflection on the right side of the coil in degree				Mean θ'=	Tan θ'
		Direct current		Reversed current				Direct current		Reversed current			
		One end θ_1	Other end θ_2	One end θ_3	Other end θ_4			One end θ_1'	Other end θ_2'	One end θ_3'	Other end θ_4'		
1.													
2.													
3.													
4.													
5.													
6.													

Graph: Plot a graph taking the distances x of the needle from the centre of the coil towards one end of the bench on X-axis by choosing a suitable scale and corresponding value of tanθ on Y-axis. The resulting curve shown in Fig.. 4.3 has two symmetric branches. Find out the points of inflexion A and B on the curve by drawing tangents

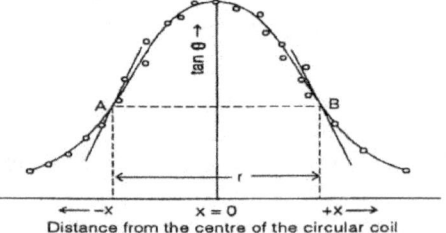

Fig. No. 4.3

on the curve. The tangent just above and just below the points of inflexion lie on opposite side of the curve. This situation will not occur for any other point on the curve.

Calculations:

1. The radius of the coil (AB) as measured from X-tanθ graph = ………….. cm
2. The radius of the coil from the measurement of circumference

$$= \frac{\text{ıference}}{} = \text{…………cm}$$

3. The percentage error in the experimental value is calculated by the following formula

    Percentage error $= \dfrac{\text{·d value} - \text{Calculated value}}{\text{·d value}} \times 100$

$$= \text{…..}\%.$$

Result:

1. The attached graph shows the variation of magnetic field along the axis of a circular coil carrying current.

2. The radius of the coil as measured from X-tanθ graph = ……..cm

Standard Result: The actual radius of the coil = …………cm.

Precautions and Sources of Error:

1. The plane of the coil should be carefully set in the magnetic meridian and the centre of the compass box should always lie on the axis of the coil.
2. There should no magnetic substance or current carrying conductors in the neighbourhood of the instrument, otherwise actual reading will be considerably affected.
3. The current flowing in the circuit should be constant and of such a value as to produce a deflection of about 70^0 when the magnetometer is at the centre of the coil.

4. While taking deflections, there should be no parallox between the pointer and its image.
5. There should be no friction between needle and its pivot.
6. X-tanθ curve should be smooth and the positions of the points of inflexion should be carefully find out.

Viva-Voce

Q.1. What are you doing?

Ans. Sir/Madam, I am studying the variation of magnetic field with distance along the axis of a circular coil carrying current.

Q.2. What do you mean by magnetic effect of current?

Ans. When a current flows in a conductor, a magnetic field is produced around it, this is called magnetic effect of current.

Q.3. What is the nature of the material of the circular coil?

Ans. The coil is made up of an insulated copper wire wound on a circular frame of non-magnetic material.

Q.4. What is the direction of field at a point on the axis of the coil?

Ans. The direction of the field is along the axis of the coil.

Q.5. What is the direction of magnetic field at the centre of the coil ?

Ans. The direction of magnetic field at the centre of the coil is normal to the plane of the coil. If the current in the face of the coil facing the observer flow in the clockwise direction then that face attains south polarity and direction of the field will be directed away from the observer. Conversely, if the current flows in the anticlockwise direction in the face facing the observer, then the field will be directed towards the obsrever.

Q.6. Why a small magnetic needle is used?

Ans. A small magnetic needle ensures that the two magnetic fields are uniform in the region surrounding the centre of the coil in which it is deflecting. The horizontal component of earth's magnetic field H is uniform over a large region but the field produced by the circular coil along its axis on passing current through it depends upon the distance of the point from the centre and thus, no-uniform, magnetic field produced by the coil F can not be uniform.

Q.7. Why the Stewart Gee type galvanometer is called tangent galvanometer?

Ans. It is called tangent galvanometer because the working of this galvanometer is based upon tangent law.

Q.8. What is Helmholtz galvanometer?

Ans. Helmholtz galvanometer is the modified form of a tangent galvanometer in which two identical vertical flat circular coils placed coaxially at a distance equal to the radius of the either coil, in Fig.4.4. Near the point on the common axis mid way

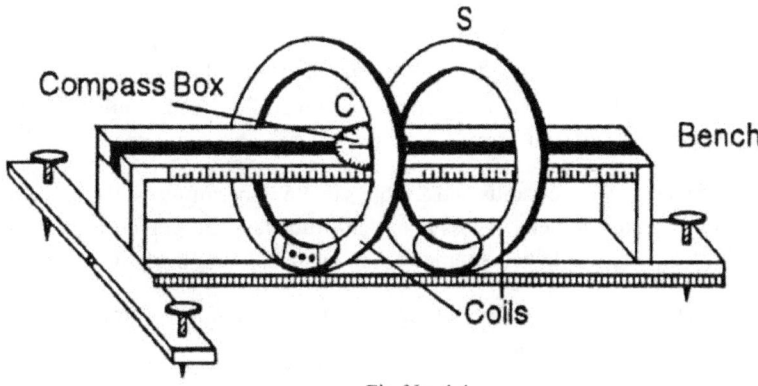

Fig. No. 4.4

between them the magnetic field is very nearly constant over an appreciable region. Therefore, the magnetic needle of the magnetic compass box placed mid-way between the two coils rotates in uniform fields. Due to this reason, a Helmholtz galvanometer is more sensitive than tangent galvanometer.

Q.9. What is the use of commutator in your experiment?

Ans. Commutator is used to reverse the direction of current in an electric circuit.

Q.10. What is point of inflexion?

Ans. The point of inflexion is the point on the x-tanθ curve where the continuous curve changes its direction of curvature or the points at which the curve changes its sign are called points of inflexion.

Q.11. What is the practical utility of finding the variation of magnetic field along the axis of a circular coil ?

Ans. The study of variation of magnetic field reveals that existence of two points of inflexion at which the rate of change of magnetic field with distance (dF/dx) is constant. This property has been used in the construction of Helmholtz galvanometer.

Q.12. What are the advantage of a Helmholtz galvanometer over an ordinary tangent galvanometer ?

Ans. These are as follows:

(i) In the ordinary tangent galvanometer the field is uniform only at the centre of the coil, while in Helmholtz galvanometer the resultant field in the entire region between the two coils is almost uniform.

(ii) For the coil of same radius and for the same number of turns, the reduction factor of Helmholtz galvanometer is less than that of the tangent galvanometer. Hence, the sensitivity of Helmholtz galvanometer is more than that of the tangent galvanometer.

(iii). For the same current, Helmholtz galvanometer produces greater resultant field and hence greater deflection.

Therefore, Helmholtz galvanometer is superior, more sensitive and accurate than tangent galvanometer.

Q.13. What is tangent law?

Ans. In two mutually perpendicular uniform magnetic fields F and H, a magnetic needle is equilibrium, makes an angle θ with the horizontal component of earth's magnetic field H, such that $F = H \tan\theta$. This is called tangent law.

Q.14. What do the number 2, 50 and 500 at the base of the apparatus indicate?

Ans. The numbers 2, 50 and 500 indicate the number of turns of the coil. The two turn coil has minimum resistance and is used to measure heavy currents. The 50 turn coil has intermediate resistance and is used to measure intermediate currents. The 500 turns coil has the maximum resistance and is used to measure weak currents.

Q.15. What is magnetic meridian?

Ans. A vertical plane passing through the axis of a magnetic needle suspended freely through its centre of gravity and rest under earth's field is called the magnetic meridian.

Experiment No. 5

Object: To determine the wavelength of spectral lines using plane transmission diffraction grating.

Apparatus required: A spectrometer, a diffraction grating, mercury lamp, prism, spirit level,reading lens, and an electric lamp.

Description of Apparatus: The entire experimental set up is divided into two part, viz.

1. Plane Transmission Diffraction Grating, and
2. Spectrometer.

1. Plane Transmission Diffraction Grating: A plane transmission diffraction grating is an arrangement consisting of large number of close, parallel, straight, transparent equidistant slits, each of equal width 'a' with neighbouring slit being separated by an opaque region of width 'b'. A grating is made by drawing a series of very fine, equidistant and parallel lines on an optically plane glass plate by means of a fine diamond pen. The light cannot pass through the lines drawn by the diamond but get scattered in different directions. The spacing between the lines is transparent to the light and act as a slit when light falls on it. There are about 15000 lines

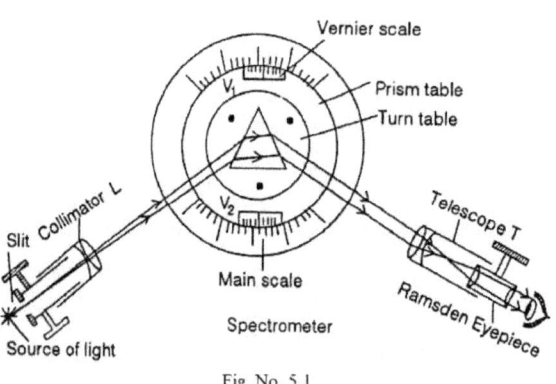

Fig. No. 5.1

per inch in such a grating to produce a diffraction of visible light. The spacing (a+b) between the adjacent slits is called diffraction element or grating element.

2. Spectrometer: An optical instrument, spectrometer is used for the study of spectra produced by prisms and gratings. A spectrometer has three main parts

(i)   Collimator,
(ii)   Graduated Prism Table, and
(iii)   Telescope (Fig. 5.1)

(i) Collimator: It consists of a long metallic tube with an achromatic converging lens L at one end and an adjustable slit at the other (which faces the source of light). The slit is fixed at the end of another smaller tube that can be slided in the long tube, by rack and pinion arrangement. The slit may be adjusted, in the focal plane of the lens L for getting a pencil of parallel rays from the collimator. The collimator tube is fixed to the base of the instrument by means of two clamps in a metal stand provided with two screws for adjusting the inclination of the axis of the collimator.

(ii) Graduated Prism Table: The prism table is a circular horizontal base on which the prism or grating can be placed. There are few concentric circles and few parallel straight lines joining two of the levelling screws on the prism table. These circles and straight lines help in placing the prism or grating on it. The base or table can be raised or lowered and can be levelled by means of three levelling screws. Prism table can be rotated about the vertical axis passing through its center. The angular position of the table can be read with the help of two verniers V_1 and V_2 (Fig. 5.1) attached to it at 180° apart and moving over graduated circular scale. At the bottom of the table a clamping screw and a tangent screw are provided for clamping and for fine setting. The parallel beam of light emerging from the collimator strikes the prism or grating resting on it.

(iii) Telescope: Like collimator, telescope also consists of a long metallic tube with an achromatic objective lens at one end (which faces the slit) and a short coaxial tube at the other end. At the one end of the short coaxial tube cross-wires are fitted and at the other end a Ramsden's eye-piece is adjusted. The eye-piece tube can move in and out with the help of rack and pinion arrangement. The telescope is so mounted that it can be rotated about the axis of the prism table and its angular position can be read on a graduated circular table on which it is attached. Telescope is provided with levelling screws, as well as with clamping and tangent screws for fine setting. The dispersed or reflected light from the prism or diffracted from grating is received in the telescope.

Formula used: The wavelength λ of any spectral line in a white light diffraction pattern obtained by a plane transmission diffraction grating of grating element (a+b) is given by

$$\lambda = (a+b)\text{----}$$

Where $\jmath 1$ = angle of diffraction in the order of the spectrum n.

The grating element, $(a+b) = \dfrac{\text{------------------------------------}}{\text{ımbe· of line pe} \jmath \text{ inci onthe ɡra:ing(N)} \quad (\ )}$

$$\therefore \quad (a+b) = \text{------} = \ldots.\text{ cm}$$

Procedure:

1. The spectrometer is placed in front of the sodium vapour lamp and its position is so adjusted that the collimator slit is illuminated with sodium light which is confirmed by viewing the slit through the collimator.
2. Turn the telescope to either side of the direct or white image of the slit to get first order spectral lines of different colours in field of view of the eye piece.
3. Now clamp the telescope and by its tangential screw set the vertical cross-wire on different spectral lines (starting from violet,green, yellow to red). For this the telescope is moved slowly by its tangential screw till the vertical cross-wire, one by one coincides with different spectral lines. Each time the reading of both verniers are recorded in the observation table.
4. Now unclamped the telescope and move it to the right of the zero order image of the slit to get the first order spectrum of the right side in the field of view of the eye-piece. Clamp the telescope again and adjust the vertical cross-wire on different spectral lines trun by turn, by tangent screw as before. Again, the reading of both the vernier scales are noted in each case with the help of reading lens and electric lamp.
5. Find the difference in the readings of same vernier for each colour separately. This gives twice the angle of diffraction for different colours in the first order spectrum (n=1).
6. Repeat the same procedure (steps 2,3,4 and 5) for second order spectrum (if visible) on both sides of zeo order image of the slit and record the readings of

verniers as before. Now calculate the angle of diffraction for the spectral lines of second order.

7. The number of lines per inch (N) ruled on the grating surface is also recorded from the grating itself.

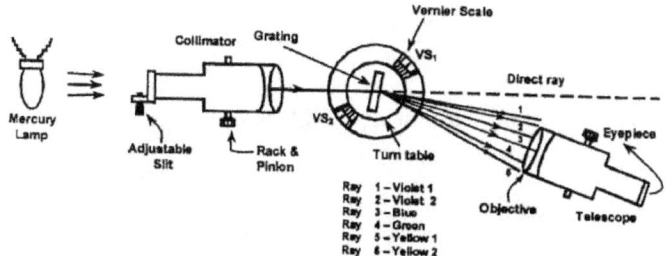

Spectrometer Arrangement – Oblique Incidence

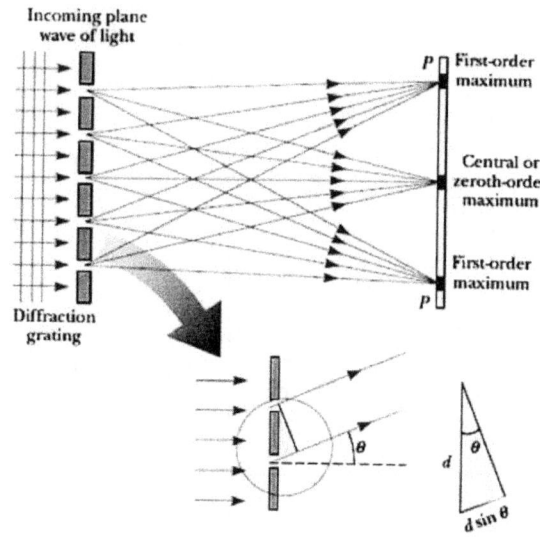

Fig. 5.2

Fig. 5.2 shows multiple spectra produced by diffraction grating.

Observations:

(1) Determination of grating element

Grating element, $a+b = \dfrac{}{\text{r of lines per inch (N)} \quad (\ )} = \ldots\ldots\text{cm}$

(2) Least count of each vernier of the spectrometer

$= \dfrac{\text{f one division on main scale} \quad \ldots}{\text{number of division on vernier scale} \quad \ldots\ldots} = \dfrac{\ldots\ldots}{\ldots\ldots} = \ldots\ldots\text{min.}$

$= \ldots\ldots\text{degree}$

(3) **Table A: For the measurement of angle of diffraction ( )**

S. No.	Order of the spectrum	Colour of the spectral line	Kind of vernier	Reading of the telescope when set on the spectral lines of different colours to								Difference of two reading of same vernier 2 $=(a{\sim}b)$	Mean value of (degrees)	sin
				Left of the direct image of the slit			Right of the direct image of the slit						
				Main scale reading (x) degrees	Vernier scale reading (nxL.c)	Total (a) (x+nxLc)	Main scale reading X	Vernier scale reading (n*L.c)	Total (b) (X+nxl.C)				
1.	First order spectrum	1. Violet1 (deep)	V1 V2										
		2.Green (deep)	V1 V2										
		3. Yellow 1	V1 V2										
		4. Red 1	V1 V2										
2.	Second order spectrum	1. Violet 1 (deep)	V1 V2										
		2.Green (deep)	V1 V2										
		3. Yellow 1	V1 V2										
		4. Red 1	V1 V2										

Calculation:

The wavelength of various spectral lines in n^{th} order are calculated by the relation

$$\lambda = \frac{(a\theta \quad)}{\underline{\qquad\qquad}}$$

The grating element, $(a+b) = \frac{2.54}{\underline{\qquad}} = \underline{\quad} = \ldots\ldots cm.$

(i)   For the first spectrum n = 1, therefore,

$$\lambda = (a+b)\sin$$

$$\lambda_{violet} = \ldots\ldots A^{\circ}$$

$$\lambda_{green} = \ldots\ldots A^{\circ}$$

$$\lambda_{yellow} = \ldots\ldots A^{\circ}$$

and $\lambda_{red} = \ldots\ldots A^{\circ}$

(ii)   For the second order spectrum n = 2,

$$\lambda = \frac{(a\theta \quad)}{\underline{\qquad\qquad}}$$

$$\lambda_{violet} = \ldots\ldots A^{\circ}$$

$$\lambda_{green} = \ldots\ldots A^{\circ}$$

$$\lambda_{yellow} = \ldots\ldots A^{\circ}$$

and $\lambda_{red} = \ldots\ldots A^{\circ}$

Mean value of $\lambda_{violet} = \ldots\ldots A^{\circ}$

Mean value of $\lambda_{green} = \ldots\ldots A^{\circ}$

Mean value of $\lambda_{yellow} = \ldots\ldots A^{\circ}$

and   Mean value of $\lambda_{red} = \ldots\ldots A^{\circ}$

Percentage Error: The percentage error in the experimental value is calculated by the following formula

$$\text{Percentage error} = \frac{\text{'d value} - \text{Calculated value}}{\text{'d value}} \times 100\% \times 100\% = \dots\%$$

Result: The experimental and standard values of d for spectral lines of various colours along with percentage erro in tablar form are given below:

S.No.	1	2	3	4
Colour of the spectral line	Violet I (deep)	Green (deep)	Yellow, I	Red, I
Experimental value λ (A°)	.... A°	.... A°	.... A°	.... A°
Standard value λ (A°)	4047 A°	4916 A°	5770A°	6234A°
Percentage Error (%)	....	....	....	....

Precautions and Sources of Error:

1. The slit should be made as narrow as possible
2. The grating should not be rubbed or touched by fingers
3. It should be insure that the ruled surface of the grating must face the telescope
4. Grating must be set exactly normal to the incident light
5. Telescope should be rotated in the same direction.
6. While taking observations, the prism table and telescope must keep clamped.
7. To set the cross wires on spectral lines of various colours tangent screw must be used.

Viva-Voce

Q.1. What are you doing?

Ans. Sir/Madam: I am determining the wavelength of various spectral lines using plane transmission diffraction grating.

Q.2. What is a diffraction grating?

Ans. It is an optically flat glass plate on which large number of equidistant parallel lines are ruled by a fine diamond pen. The space between the successive lines are transparent while the lines drawn are opaque to light.

Q.3. How many order of spectral are seen here?

Ans. Only, first and second order are observed.

Q.4. What is the main difference between the spectrum obtained by grating and that due to prism?

Ans. In grating spectrum, red colour is deviated most and violet least. The sequence of the colours in grating spectrum is reverse than that of prism spectrum (VIBGYOR).

Q.5. If the number of lines on your grating be doubled, what will happen to the grating element?

Ans. If the number of lines on grating be doubled, its grating element will be reduced to half of its previous value.

Q.6. What are the essential condition of obtaining pure spectrum?

Ans. The essential conditions for obtaining pure spectrum are as follows:

   (i)   The prism should be placed exactly in the position of minimum deviation.
   (ii) A narrow and parallel beam of light should fall on the prism.
   (iii)A convex lens of suitable focal length should be used to focus the spectrum.
   (iv)To see the virtual enlarged spectrum another convex lens should be used.

Q.7. What is normal spectrum?

Ans. Normal spectrum is one, in which the angular reperation is directly proportional to the difference in the wavelength.

Q.8. What is angle of deviation?

Ans: The angle between the incident ray and the emergent ray emerging from the prism is called angle of deviation.

Q.9. What do you mean by dispersion of light?

Ans. When a ray of white light falls on a prism, it splits up into its constituent colours. This phenomenon of splitting of light into its constituent colour is called dispersion.

Q.10. Why a white light on passing through prism disperses into its constituent colours?

Ans. It is because of the fact that the velocity of different colours in material medium is different. Hence, the refractive index μ of a material is different for different colours of light. Since the angle of deviation δ depends on the refractive index of the material, the rays of different colours emerges in different directions. In the glass the speed of violet light is minimum while that of red light is maximum. Therefore, the refractive index of glass is maximum for the violet light and minimum for red light. Hence, tha angle of deviation for the violet light will be greater than the angle of deviation for the red light. When white light enters a glass prism the ray of violet light bends maximum towards the base while red bends least.

Q.11. On what factors does the refractive index of a material depend?

Ans. Refractive index of a material depends upon:

(i)   Nature of the material,
(ii)   Nature of the medium,
(iii)   Wavelength of light and
(iv)   Temperature.

Q.12. On what factors does the dispersive power depends?

Ans. The dispersive power of a grating depends on the following factors:

(i)   The dispersive power of a grating is directly proportional to the order of spectrum (n), that is, higher orders are dispersed more than the lower orders.

(ii)   The dispersive power of a grating is inversly proportional to the grating element, that is smallest the grating element (a+b), more is the dispersive power.

(iii)   The dispersive power of a grating is inversly proportional to the cosine of the angle of diffrection, that is, the larger the angle of diffraction more is dispersive power, that is, the dispersion is more in red region than in violet region.

Experiment No. 6

Object: To determine the specific resistance of a given wire using Carey Foster's Bridge.

Apparatus required: Carey Foster's bridge, resistance box, Galvanometer, Battery, Eliminator, plug key, rheostat, connection wires and jockey.

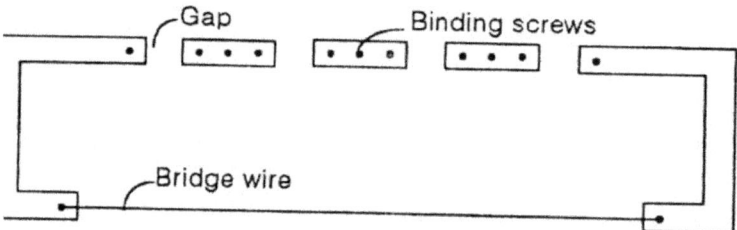

Fig. No. 6.1

Description of Apparatus: Carey Foster bridge is the modified form of a meter bridge having two addition gaps as shown in Fig. 6.1. It is used to measure very small resistances. A Carey Foster bridge consists of a meter long wire of uniform cross-section stretched along a wooden board and runs parallel to a meter scale. The wire is of the material of high specific resistance and low temperature coefficient, such as ureka or manganin. The ends of the wire are soldered to two L shaped bass strips. The three brass running parallel to the wire are firmly attached on the board between the two L shaped strips as shown in these five strips constitute four gaps. Terminals with binding screws provided on each strip for making necessary

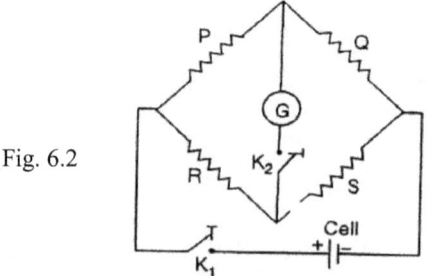

Fig. 6.2

connections. The working of Carey Foster bridge depends on the principle of Wheatston's bridge, as is clear from Fig. 6.2.

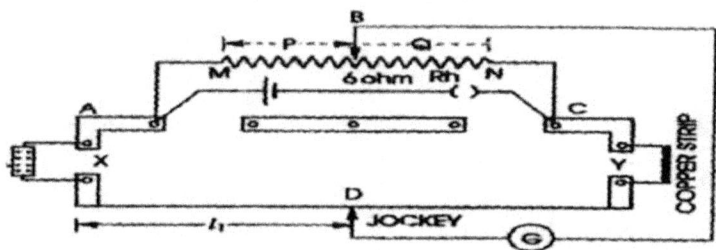

Fig No. 6.3

The circuit arrangements for the determination of specific resistance of the material of a wire is shown in Fig.No. 6.3. In this arrangement two nearly equal and small resistances P and Q are connected in the inner gap of the bridge. These resistances serve as a ratio arms of the Wheatstone bridge. Of the two outer gaps one is short circuited by connecting a thick copper strip across its ends, and to the other outer gap a fractional resistance box is connected. A leclanche cell in series with a key K is connected to the terminal D and E of the two middle strips. One end of the sensitive galvanometer G is connected to the point O of the central strip and other to the jockey J which slides over the bridge wire AB. For the measurement of resistance of a wire, the copper strip is replaced by the wire whose resistance is to be measured.

Before measuring the actual diameter or radius of the experimental wire with the help of screw gauge, we first measure the pitch, least count and zero error of the screw gauge in the following manner.

To determine the pitch of the screw gauge rotate the cap of the screw till the zero mark of the circular scale coincide with the reference line. Note this reading on the linear scale. Now give four or five complete rotations to the circular scale and note the reading on the linear scale. The difference of the two reading on the linear scale gives the linear distance moved when a certain number of rotations to the circular scale has been given. Calculate the pitch of the screw gauge by the formula

$$\text{Pitch} = \frac{\text{tance moved along the linear scale}}{\text{mplete rotations given to the drdar scale}}$$

Least count of the screw gauge is obtained by the relation,

$$\text{Least count} = \frac{}{\text{).of division on the circular scale}}$$

Formula used: The resistance per unit length(e) of the bridge wire is given by

$$X-Y = {}_{.1}(L_2-L_1)$$

where X = Known resistance,
                 Y = Unknown resistance and
                 L_1 & L_2 = Length of balance point on bridge wire before and after the interchanging the resistance, R.

                 If $Y = 0$ and $X = R$, then

$$\rho = \underline{\quad\quad}$$

                 So, the specific resistance $K = \underline{\quad\quad}$

                 where r = radius of the wire

                 and L = length of the given wire.

Procedure:

1. Connected Ohm box in the left gap and a thick copper strip in the right gap of a C. F. bridge. Lower terminals of rehostat MN are connected to A and C while the sliding contact B to a jockey D through the galvanometer G. Better eliminator with key is also connected between A and C (Fig. 6.3).

2. This sliding contact B of the rheostat MN is kept as its mid value middle so that resistance P and Q may be nearly equal.

3. Introduce some resistances R in the decimal Ohm box in left gap, X, and by sliding Jockey null point is obtained. The distance of the null point II from left end is noted.

4. Now the position of resistance box and copper strip are interchanged. Again the balance is obtained for same resistance in decimal Ohm box. Let the distance of the null point from the same end i.e. left end be L_2.

5. Calculate the value of ρ by following formula

$$\rho = \frac{R}{L \quad L}$$

6. To measure the unknown resistance Y, remove the thick copper strip from the right gap. Introduce a suitable resistance in decimal Ohm box such that balance is obtained at a distance L_1 on the bridge wire. Interchange the positions of resistance box and again obtain the distance of null point (L_2).

Then, $Y = X-_1 (L \quad L)$

Where X ia a known resistnance which is introduced in the resistance box. Take different sets by changing the value of resistance in decimal resistance box.

Observations:

(1) Table A: For determination of :

	Resistance introduced in Ohms box (x)	Position of null point with decimal Ohm box in		L_2-L_1	$\dfrac{\quad}{_1} = \dfrac{\quad}{\quad}$	Mean ( ). Ohm/ cm
		Left gap L_1	Right gap L_2			
1.						
2.						
3.						
4.						

(2) Table B: For the determination of resistance wire

S.No.	Resistance introduced in ohms box (x)	Position of null point with decimal ohm box		L_2'-L_1'	$Y = X - \rho(L_2' - L_1')$	Mean Y Ohm
		Left gap(L_1')	Right gap(L_2')			
1.						
2.						
3.						
4.						

(3) For the measurement of radius (r) of the experimental wire:

Least count of the screw gauge = $\dfrac{\rule{4cm}{0.4pt}}{\text{).of division on the circular scale}}$

$= \dfrac{....}{....} = ...$ cm.

Pitch $= \dfrac{\text{tance moved along the linear scale}}{\text{implete rotations given to the ardar scale}}$

$= \dfrac{....}{....} = ...$ cm

Therefore,

Diameter of wire = M. S. reading + total number of division of circular scale x L.C.

Radius = $\dfrac{\text{ər}}{.}$ $= ....$cm

Calculations:

    (i)   The resistance per unit length (ρ) of the bridge wire is calculated separately for each set of observations, given in table A, with the help of the formula

$$\dfrac{.}{.} = \dfrac{.}{L_2 - L_1}.$$

And then its mean value is obtained.

(ii)   The resistance (Y) of the experimental wire is calculated separately for each set of observations recorded in table B, with the help of the relation $Y = X - \rho(L^{-} . L' \textrm{)} = \ldots$ Ohm.

(iii)   By measuring the radius (r) of the experimental wire accurately, the specific resistance K of the material of the experimental wire is calculated by using the formula

$$K = \frac{\qquad}{\qquad} = \ldots \textrm{ohm-cm}$$

Standard Result: The standard value of the specific resistance for the material of the given wire (Copper) = 1.78×10^{-6} ohm cm.

Result: The specific resistance of the given wire = ... ohm cm

Percentage Error: The percentage error in the experimental value is calculated by the following formula

$$\textrm{Percentage error} = \frac{\textrm{'d value} - \textrm{Calculated value}}{\textrm{'d value}} \times 100 \times 100 \% = \ldots \%$$

Precautions and Sources of Error:

1. The connection wire should be thick and their ends should be properly cleaned with sand paper before use. All the connections to the binding terminals should be tightly made.
2. The decimal ohm box should be connected to the bridge gap by short and thick copper wires.
3. The plug key connected in the cell circuit should be closed only when observations are to be taken.
4. First the plug key of the cell circuit should be closed and then jockey should be pressed gently on the bridge wire.

5. To avoid the excessive deflection in the sensitive galvanometer, the galvanometer should be shunted by a wire of low resistivity which should be removed as one reach near the balance point.

Viva-Voce

Q.1. What are you doing?

Ans. Sir/Madam, I am determining specific resistance of a given wire by Carey Foster's Bridge.

Q.2. What do you mean by specific resistance?

Ans. The specific resistance of the material of a wire is defined as the resistance of one cm cube of the material or the resistance of one cm length of a wire whose area of cross-section is one square centimeter.

Q.2. What type of wire is used in making the resistance coils?

Ans. In the preparation of resistance coils the constantan or manganin alloys are used. The constantan(Cu 60% and Ni 40%) or manganin (Cu 84%, Ni 4%, Mn12%) wires are doubly insulated by means of silk thread.

Q.4. What is the working principle of a Carey Foster's bridge?

Ans. When the resistances connected to the outer gaps of the Carey Foster bridge are interchanged the position of the null points changes. The difference in the value of resistances (X-Y) is equal to the resistance of that much of the bridge wire [ρ(L   L)] by which the null point has shifted.

Q.5. What do you mean by the least count of a screw gauge?

Ans. Least count is the least reading which can be measured accurately with a screw gauge. It is the distance moved along the linear scale, when circular scale moves through one division.

Q.6. What precaution are you taking in regard to the use of jockey?

Ans. The jockey should not be pressed hard on the bridge wire nor should it be slipped over it, otherwise the diameter of wire may not be uniform.

Q.7. If the length or radius of a wire is increased, whether the specific resistance will change?

Ans: No, because the specific resistance is independent of length or radius of the wire, it depends only on the material of a wire.

Q.8. What is the principle on which the Carey Foster's bridge is based?

Ans: Carey Foster's bridge is based on the principle of Wheatstone's bridge.

Q.9. What is a resistance box?

Ans: A resistance box has a number of resistance coils of different resistances, all are connected in series and mounted inside a box in Fig. 6.4.

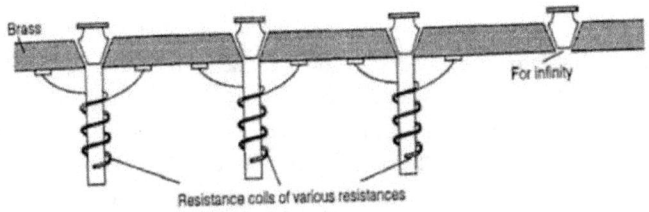

Fig No. 6.4

Q.10. How the radius of the resistance wire is determined?

Ans. The radius of the resistance wire is determined by means of a screw gauge in two mutually perpendicular directions at a number of places and then mean is taken.

Q.11. What do you mean by the least count of a screw gauge?

Ans: Least count is the least reading which can be measured accurately with a screw gauge. It is the distance moved along the linear scale, when circular scale moves through one division.

Experiment No. 7

Object: To determine the specific rotation of cane sugar solution using half shade polarimeter.

Apparatus required: Laurent's half shade polarimeter, Sodium lamp, Sugar, Balance weight box, Measuring cylinder, Beaker, Eyepiece, Glass rod, Reading lens, funnel and Reading lamp.

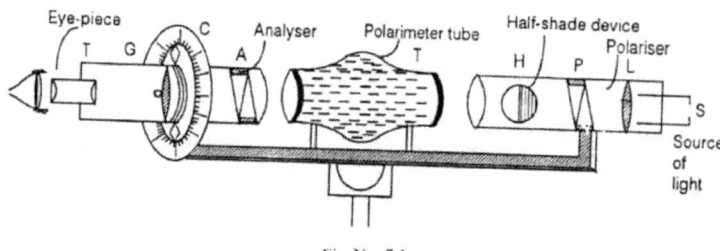

Fig. No. 7.1

Description of Apparatus: Fig. 7.1 shows the schematic arrangement of Laurent's half shade polarimeter. It consists of two nicols P and A mounted in separate brass tubes placed some distance apart and are capable of rotation about a common axis. A glass tube T having a larger diameter in middle contains the active solution under examination. Two ends of the glass tube are covered by flat and parallel glasss plates or caps. It is mounted between the polariser P and analyser A on a rigid iron base.

Monochromatic light of wavelength λ from a source S, rendered parallel by a convex lens L, falls on a nicol prism P. After passing through P the light becomes plane polarised. This plane polarised light now pass through a half shade device H and then through the solution whose specific rotation is to be determined and filled in the tube T. The transmitted light passes through analysing nicol A, which can be rotated about the direction of propagation of light. The emergent light from nicol A is viewed through a Galilean telescope G.

Action of Laurent's Half Shade Plate: The Laurent system consists of a half shade plate in two halves, one of quartz cut parallel to its optic axis and the other a matching plate of glass [Fig. 7.2(a)] so chosen as to absorb and reflect the same amount of light as the quartz plate.

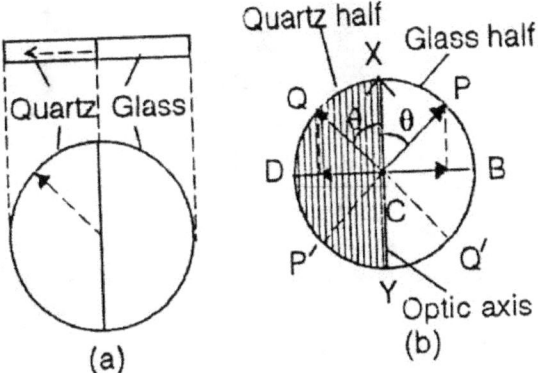

Fig No. 7.2

The quartz is a half wave plate which introduces a path difference of $\lambda/2$ between the ordinary and extra ordinary rays in transmission normal through it. Let the plane of vibration of the plane polarised light from the polariser P falls normally on half shade plate along CP [Fig. 7.2(b)]. The light passing through the glass plate remains unaffected, but that falling on the quartz plate is broken up into two components-one E-component CX parallel to the optic axis XY and the other O-component perpendicular to the optic axis, that is, along CB. As in quartz, O-component travels faster, hence, on emergence O-component will gain a phase change of th over the E-component. Thus on emergence from the quartz plate, O-component has vibrations along CD and E-component has vibrations still along CX. The emergent wave CQ is the resultant of vibrations along CD and CX. Here, angle PCX = angle. Thι=θ. Thus the angle between vibration planes of light emerging from quartz CQ and that of light emerging from glass CP is 2θ.

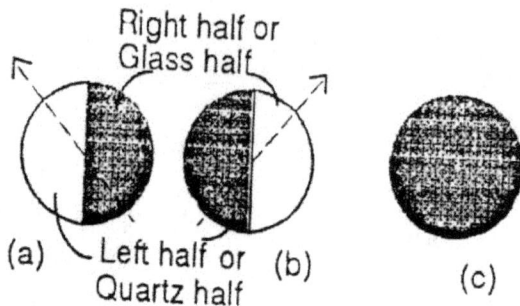

Fig. No. 7.3

Thus, there are two plane polarised beam-one emerging from glass with vibrations in the plane CP while other emerging from quartz with vibrations in the plane CQ. If the principal plane of the analysing nicol A parallel to QCQ', the light from quartz plate will pass unobstructed while from glass plate will be partly obstructed. Thus the quartz half will be brighter than the glass half [Fig. 7.3(a)]. If the principal plane of the analysing nicol A is parallel to PCP', then the light from glass plate will pass unobstructed, the light from quartz plate will be partly obstructed. Thus the right half will appear brighter as compared to the left half [Fig.7.3(b)]. But when the principal plane of analysing nicol A is parallel to the optic axis XCY, the two halves appear equally illuminated [Fig.7.3(c)].

Formula used: If r is the rotation of plane of polarisation of incident polarised beam produced by L decimeter length of the cane sugar solution or of any optically active substance with concentration in gm/c.c. then the specific rotation at a given temprature t corresponding to the wavelength λ is given by

$$ = -$$

where   a = Rotation in degree

and   l = length in dm

Procedure:

1. Take a clean dry beaker and add to it 10 gm sugar and weigh again with beaker. Calculate volume of solution to have a 10% strength as follow:

   Volume required $= m\text{——}$

   where, m is the mass of sugar.

2. Find vernier constant of the circular scale place the polarimeter so that the aperture in the front of the sodium lamp. Adjust the position of the eye-piece so that two halves of the half shade device are clearly in focus.

3. Now, clean the tube. Hold the tube in vertical position and fill it with water. Slip the second glass window gently on the tube taking care that no air bubble are left.

4. Place the tube in the polarimeter in position and cover it.

5. Rotate the analyzing nicol by rotating the circular scale till the two halves of the half shade device are equally dark in colour. In this position there will be an abrupt change in the intensity of two halves when the slight rotation on other side is given. Note reading on the scale.

6. Now fill the tube with 20% sugar solution prepared as explained above and note the reading of the scale following the same experimental procedure.

7. Repeat the experiment by turning through 180°.

Observations:

(1) For the concentration of cane sugar solution:
   Mass of the empty flask (a) = …… gm
   Mass of the flask with cane sugar (b) = ….gm
   Therefore,
   Mass of the cane sugar, m(b-a)= …. gm
   Volume of water in the measuring flask, $V = \dots\text{cm}^3$

   Concentration of the sugar solution, $c = \frac{..}{..}$ ……… $\frac{..}{..}$ = ….. gm/cc

   Room temperature t = …. °C

   Length of the glass tube l   = ….cm …. $= \frac{....}{....} = $ ….decimeter

   Wavelength of the sodium light used = 5893 Å = 5893 × 10   cm

(2) Table A: For the measurement of angle of rotation θ:

Value of one division of main scale   = ...
Total number of divisions on vernier scale = ...

Therefore,

$$\text{Least count of the analyser} = \frac{\text{f one division of main scale}}{\text{divisions on the vernier scale}}$$

$$= \frac{...}{....} = \dots \text{ degree}$$

(i)   With pure water in the glass tube:

S.No.	Clock wise			Anti clock wise			Mean
	M.S. in degree	V.S. in degree	Total	M.S. in degree	V.S. in degree	Total	a
1.							
2.							
3.							
4.							

(ii)   With cane sugar in the glass tube:

S.No.	Concentration of Solution (gm/cc)	Clock wise			Anti clock wise			Mean b=═ₑ	⋅ l= —
		M.S. in degree	V.S. in degree	Total	M.S. in degree	V.S. in degree	Total		
1.									
2.									
3.									
4.									

Calculations:

1. Concentration of the solution $c = \frac{\cdot}{\cdot} = $ ......$\frac{\cdot}{\cdot}$

    Specific rotation of cane sugar $S = \frac{x}{x} = $ ......$°/l/c\sqrt{}$ /

2. From graph

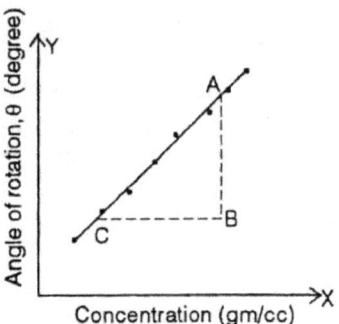

Fig. 7.4

The specific rotation can also be determined by plotting a graph between the angle of rotation θ and corresponding concentration of the solution, C [Fig 7.4]. The graph is a straight line whose slop is

$$\frac{3}{\tau} = \underline{\quad}$$

Therefore, $S = \frac{B}{C} = -(\frac{\cdot}{\cdot}) = $ ......$°/dm/gm/cc = $ ......$°/dm/kg/m^3$

Percentage Error: The percentage error in the experimental value is calculated by the following formula

Percentage error $= \dfrac{\text{d value} \sim \text{Calcu}\sim\text{ted value}}{\text{d value}} \times 100\% = $ ...%

Result - The specific rotation of cane sugar solution for sodium light of wavelength

$3 = 5983\text{Å}$ is ......$°//m/$ /

Standard Result: Standard value of the specific rotation of cane sugar is

$66.5°/dm/kg/m$.

Precautions and Sources of Error:

1. The glass tube should be properly cleaned before use.
2. The solution in the glass tube should essentially be free from air bubble.
3. The tube should be well rinsed with the solution of each concentration before filling.
4. The caps at the ends of the glass tube should not be tightly screwed after filling the tube with solution.
5. The position of the analyser should be precisely set for recording each observation.
6. The room temperature and the wavelength of light used should be mentioned in describing result.

Viva-Voce

Q.1. What are you doing?

Ans. Sir/Madam, I am determining the specific rotation of cane sugar solution with the help of Laurent's half shade polarimeter.

Q.2. What do you mean by specific rotation?

Ans. The specific rotation of an optically active substance at a given temprature for a given wavelength of light is defined as the rotation of plane of polarisation of incident polarised beam produced by one decimeter length of the substance of unit density.

If **r** is the rotation produced by 1 decimeter length of a substance, the concentration of its solution is c gm/cc., then specific rotation S at a given temprature t for a given wavelength **r** is expressed as

$$= -$$

Q.3. What is plane of polarisation and plane of vibration?

.

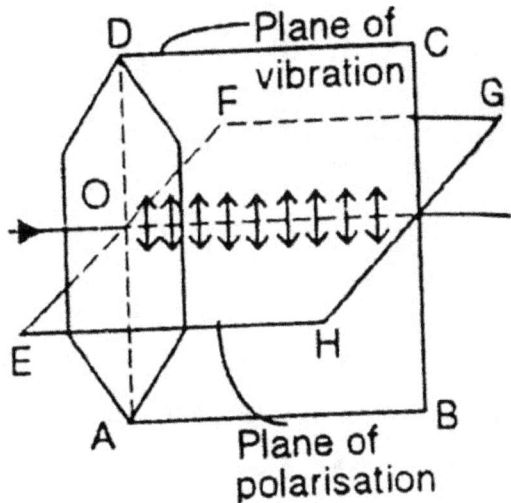

Fig.7.5

Ans. **Plane of vibration:** The plane containing the direction of vibration and the direction of propagation of light is called the plane of vibration. ABCD is the plane of vibration in Fig.7.5.

Plane of polarisation: The plane passing through the direction of prapagation and containing no vibration is called the plane of polarisation. The plane of polarisation is always perpendicular to the plane of vibration. EFGH is the plane of polarisation in Fig. 7.5.

Q.4. What is a polarimeter?

Ans. Polarimeter is a device used for the measurement of optical rotation and the angle of rotation of the plane of polarisation rotated by an optically active substance.

Q.5. What are the main parts of a polarimeter?

Ans. The main parts of a polarimeter are; a polariser, an analyser and a polarimeter tube kept between them.

Q.6. What is nicol prism?

Ans. William Nicol invented and constructed an optical device made from a calcite crystal for producing and analysing plane polarised light.

Q7. What is the unit of specific rotation?

Ans. The unit of specific rotation is degree/decimeter/gm/c.c. or degree/decimeter/kg/m^3.

Q.8. What is the value of specific rotation of cane sugar solution in water?

Ans. A specific rotation of cane sugar solution in water at 20^0C is $+66.5^0$.

Q.9. What is the significance of plus sign in the above value of specific rotation?

Ans. The plus sign indicates that the rotation is clockwise or right handed.

Q.10. What is the main difference in the working of half shade and Bi-quartz polarimeters?

Ans. The main difference in the working of half shade and Bi-quartz polarimeter is - In the half shade polarimeter, the monochromatic sodium light is used and in it two halves in eye-piece appears of different illumination, whereas in bi-quartz device, white light is used and in it two halves in eye-piece appears of different colours.

Q.11. What do you mean by optically active substance?

Ans. Certain substance have a tendency to rotate the plane of polarisation of a plane polarised light when propagated through it. Such substances are called optically active substances.

Q.12. What is the practical utility of the measurement of this specific rotation?

Ans. It is extensively used in sugar factories for the estimation of the percentage of sugar in a given solution in sugar factories. This method is also used to determine the amount of sugar present in the urine of a diabetic patient.

Q.13. What is the use of half shade device in your experiment?

Ans. A half shade device is used in the experiment to judge the accurate position of complete darkness of the field of view. The analyser is unable to detect the exact position of complete darkness. When the analyser is rotated through some angle to detect the position of complete darkness, the field of view remains practically dark even after the analysing nicol has been rotated through 5 or 6^0 near the crossed position.

Q.14. What is the construction of half shade device?

Ans. It consist of two semi-circular plates. One semi-circular plate is of ordinary glass, whereas the other is of calcite. Both are cemented together along the diameter. The quartz is a half plate, it introduces a path difference of $\lambda/2$. The thickness of the glass plate is so chosen as to absorb and reflect the same amount of light as the quartz plate.

Q.15. Why do you use sodium light with half shade device?

Ans. The half shade device in the polarimeter produces a path difference of $\lambda/2$ between ordinary and extra ordinary rays for a particular wavelength λ for which it is designed. Generally this wavelength is matched with the wavelength of sodium D line. Hence the use of sodium light is necessary.

Experiment No. 8

Object: To verify Stefan's law by electrical method.

Apparatus Required: A 6 volt battery to heat the filament of the diode, vacuum diode valve value EZ-81, D.C. voltmeter (0-10 volts), D.C. ammeter (0-1 Amp) and rheostat (100Ω). Usually all the above component are arranged in a single cabinet.

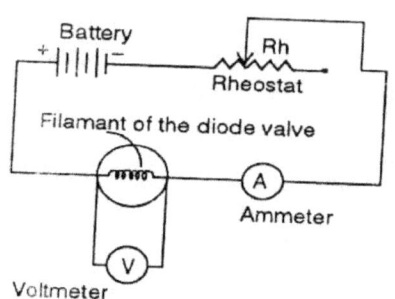

Fig. No. 8.1

Description of the Apparatus: In this experiment we use a vaccum diode EZ-81 which has a cylindrical cathode made of nickel. The tungsten heater filament is closely fitted inside the cathode sleeve. A mixture of barium and strontium oxides is sprayed over the outer surface of the nickel sleeve from which thermionic emission takes place. Since the cathode sleeve and the heater filament are in close physical contact we take the temperature of the filament. Cathode of the diode is heated by passing electric current through the tungsten heater filament. To connect the voltmeter across the valve, the two wires are soldered to the base point of the valve. A battery, rheostat Rh and an ammeter are connected in series with the filament of the valve as shown in Fig. 8.1.

Theory and Formula Used:

If E be the total energy radiated per second from a unit surface area of a black body at a temperature T surrounded by another body at temperature T_0, then by Stefan's Law

$$E = \sigma(T^4 - T_0^4) \qquad \ldots\ldots\ldots\ldots(1)$$

where, σ is the Stefan's constant. For the bodies other than the black body, the similar relation for the power emitted by a body at temperature T surrounded by another body at temperature given by

$$P = \cdot \ ' \left(\begin{smallmatrix}\alpha\end{smallmatrix}\right) - \ .. \ ) \qquad \ldots\ldots\ldots\ldots\ldots(2)$$

Where, C is some constant which depends on the material and area of the body and α is a power very close to 4. Further

$$P = - \ \tfrac{\iota_{\dot{c}}}{\cdot}(1 - - \cdot)$$

If $T \gg T_0$, then the above relation reduces to

$$P = \ \ldots \ldots \qquad \ldots\ldots\ldots\ldots\ldots(3)$$

Taking logarithm of both sides, we get

$$\mathrm{Log}_{10}P = \alpha \ \log_{10}T + \log_{10}C \qquad \ldots\ldots\ldots\ldots(4)$$

Eq.. (4) is a form of straight line equation like, $Y = mx + C$. Therefore, a graph between $\log_{10}P$ and $\log_{10}T$ should be a straight line whose slope gives α. If α is a approximately equal to 4, then Stefan's law is verified. Hence, to verify Stefan's law, we have to measure the following two quantities:

1. **Power P radiated by the body:** In this electrical method tungsten filament of the vacuum, diode is used as the radiating body. In the steady state, if we neglect the power loss due to conduction and convection, the electrical power should be equal to the power P radiated by the body.

2. **Temperature T of the radiating body:** The temperature of the radiating body or filament is determined by using well known resistance-temperature relation expressed as

$$\frac{\text{\text---}}{\text{\text253}} = (\text{\textemdash}) \qquad \ldots\ldots\ldots\ldots\ldots(1)$$

Therefore, the operating temperature of the filament is determined by measuring its electrical resistance.

For the use of above formula we need filament resistance R_{273K} at 273K or 0^{0}C which can be measured by determining the resistance at room temperature, that is, R_{300K} and the temperature coefficient of resistance, $\alpha = 0.0053$ K^{-1} [$R_t = R_0(1+\alpha t)$]. $R_{300\ k}$ can be determined by measuring the filament resistance at low filament voltage(<1 volt) and extrapolating their resistance to zero volt. For the EZ-81 vacuum tube R_{300} is 0.6. Therfore, it is convenient to use a graph between T and R_T/R_{300} instead of using relation (1) for the determination of filament temperature for tungsten from the published results shown in Table A.

Table A (For Tungsten)

S.No.	R_T/R_{300}	T(K)
1.	1	300
2.	4	920
3.	6	1300
4.	8	1645
5.	10	1990

The above result are not valid if the filament deteriorates due to prolonged use or due to oxidation of the surface of the tungsten filament. For the electrical insulation, the filament placed inside the cathode sleeve is covered with a thin coating of plaster of paris and diode is evacuated to high degree of vacuum.

Procedure:

1. Make the electrical connection [Fig.8.1] and main supply is switched on. Now keep the current at zero position by adjusting the current control knob at minimum.
2. Now current I is increased and apply some filament voltage by adjusting current control knob one by one at 0.2V, 0.4V, 0.6V......Volts etc. and measure the corresponding filament current I_f in the ammeter steady state is reached. For steady state, wait for 3 or 4 minutes before recording the observations after adjusting the filament voltage V_f'
3. Repeat the experiment for sufficient number of set of observations so that graph can be plotted.

Obversations:

(1) Table B

S.No.	Filament voltage V_f (volts)	Filament current I_f (mA)	Power radiated $P = V_f I_f$ (watt)	Filament resistane $R_T = V_f/I_f$ (ohm)	R_T/R_{300K} (For EZ 81 tube, $R_{300K} = 0.6$)	T(K)	$\log_{10}P$	$\log_{10}T$
1.	0.2							
2.	0.4							
3.	0.6							
4.	0.8							
5.	1.0							
6.	2.0							

Calculations:

1. Power radiated by the tungsten filament for each value of filament voltage V_f and filament current I_f is calculated as, $P = V_f I_f$

2. Filament resistance R_T is calculated for each value of filament voltage V_f and filament current I_f as $R_T = V_f/I_f$ and find the value of R_T/R_{300K} (use $R_{300K} = 0.6$ for EZ-81 diode).

3. Now a graph is plotted between the given values of T(K) and R_T/R_{300K} for tungsten with the help of Table A. From this graph (Fig. 8.2) for each experimentally calculated value of R_T/R_{300K} find corresponding value of temperature T(K).

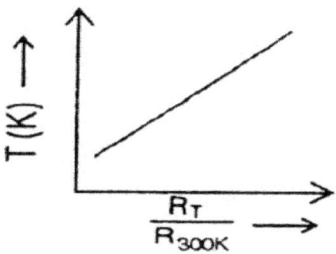

Fig. No. 8.2

4. Now plot another graph taking $\log_{10}T$ along X-axis and corresponding $\log_{10}P$ along Y-axis. The graph will be straight line (Fig.8.3). Find its slope which be about 4. That is, the slope of the curve.

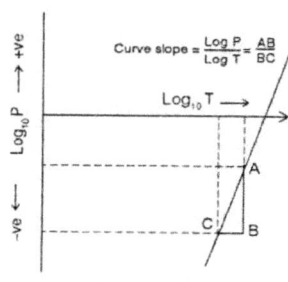

$$\frac{AB}{DC} = \frac{}{} = 4$$

Fig. No. 8.3

Result:

The graph between $\log_{10}P$ and $\log_{10}T$ is a straight line and the slope of a straight line is about 4(.....). Hence, Stefan's law is verified within the experimental error.

Precautions and Sources of Error:

1. To get accurate resistance at a particular temperature the filament volatge V_f and filament current I_f should be read every time after achieving a steady state or the time difference between each observation should be about 3 or 4 minutes.

2. The slope of the straight line should be determined as accurate as possible.

Viva-Voce

Q.1. What are you doing?

Ans. Sir/Madam, I am verifying Stefan's law by electrical method.

Q.2. What is Stefan's law?

Ans. $E \propto T^4$ or $E = \sigma r T^4$ where σ is called Stefan's constant, E is the energy rediated per second by unit area of a perfectly black body and T is the absolute temperature of the surface of the body.

Q.3. What is Kirchoff's law of radiation ?

Ans. At a given temperature the ratio of the emissive power to the absorptive power for a particular wavelength is constant and it is equal to the emissive power of a perfectly black body at the same temperature for the same wavelength.

Q.4. What is a black body ?

Ans. A black body absorbs all the incident radiation irrespective of frequency.

Q.5. What is the importance of Stefan's constant ?

Ans. Stefan's constant is used to determine the temperature of the heavenly bodies like sun.

Q.6. What is the value of Stefan's constant?

Ans. The value of Stefan's consatnt 6 is 5.67×10^{-5} erg/(sec-cm^2-K^4) in C.G.S. system and 5.67×10^{-8} joule/(sec-m^2-K^4) or 5.67×10^{-8} watt/m^2-K^4 in SI system.

Q.7. What would be the form of Stefan's law for the non-perfectly black bodies?

Ans. If the body is not perfectly black, then the total power emitted by a body at temperature T surrounded by another body at temperature T_0 is given as

$$P = C(T^\alpha - T_0^\alpha)$$

where α is a power quite closed to 4 and C is some constant depending on the material and area of the body.

$$P = CT^\alpha(1 -- \cdot).$$

$$\text{If } T >> T_0, \text{ then } P = CT^\alpha$$

Q.8. Is Ohm's law valid in your experiment?

Ans. No, because the resistance of the filament of the vacuum diode, that is, the ratio V/I begins to increase with the increase in potential difference V. It is because of the fact that when the current in the filament increase enough, then its temperature becomes quite high and therefore its resistance increases. In fact the circuit having vacuum diode is called non-ohmic circuit.

Q.9. What is radiating body in your experiment?

Ans: Radiating body in our experiment is tungsten filament.

Q.10. What is the basic difference between the perfect black body and non- perfect black body?

Ans: The emissivity of a perfectly black body is unity, whereas it is less than one for non-perfectly black bodies.

Q.11. What happens to the resistance if the temperature is raised?

Ans: Resistance increases with the increase of temperature.

Experiment No. 9

Object: To calibrate the given ammeter and voltmeter with the help of a potentiometer.

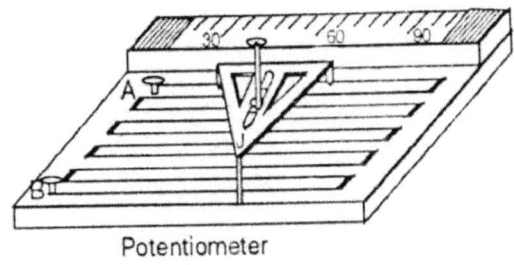

Potentiometer

Fig No. 9.1

Apparatus required: A potentiometer, two rheostats, two storage batteries,, voltmeter, a sensitive galvanometer, a standard cadmium cell, standard one Ohm resistance coil, ammeter, one two way key, two one way keys and connection wire.

Description of Apparatus: A ten or twelve wire potentiometer consists of a long uniform wire of a material having high specific resistance and low temperature coefficient of resistance such as manganin or constantan, stretched between two terminal on a wooden board. In this type of potentiometer instead of one wire. 10 or 12 wire each of length one meter are connected in series with the help of metallic strips along a meter scale fixed on the board as shown in Fig. 9.1

The entire circuit is arranged into two parts: **Main circuit** and **Auxiliary circuit.** The main circuit of a potentiometer is set up by connecting the main battery B_1, a rheostat (110 Ω range) Rh_1 and a key K_1 in series between the two extreme ends A and B of the potentiometer. The positive terminal of the storage battery B_1 is connected to the zero and A of the potentiometer wire. The auxiliary circuit is made by connecting the other battery B_2 in series with a rheostate (50 Ω range) Rh_2, a standard one ohm resistance coil, an ammeter A to be calibrated through a key K_2. A voltmeter which is to be calibrated is connected across the one ohm coil as shown in Fig. 9.2. The positive terminal of the standard cadmium cell and the higher potential terminal of the standard one ohm resistance coil of the auxiliary circuit are connected to the zero end A of the potentiometer wire.

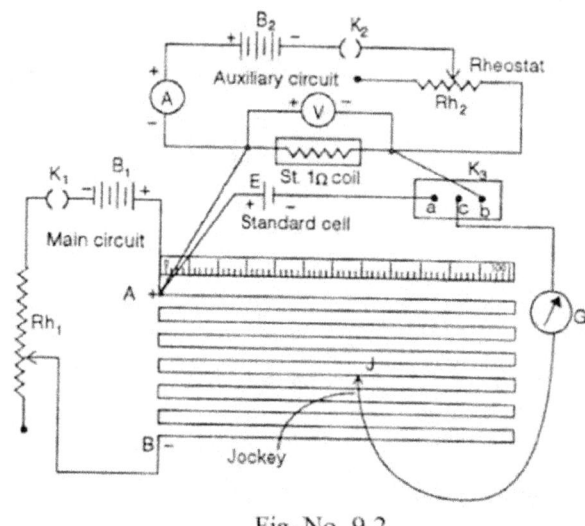

Fig. No. 9.2

The negative terminal of the standard cadmium cell and the lower potential terminal of the standard 1-ohm coil are connected to the two side terminal a and b of two-way key K_3 whose central terminal c is connected to the sliding jockey J through a galvanometer G.

Formula used:

Error in Voltmeter reading V' - V = [V}   ]

Error in Ammeter reading I' – I = [E- – I]

and   V' = I'= Kl_2

where,

V' = potential difference between two points by potentiometer,

V = Potential difference between the same two points by voltmeter,

I' = Current equal to voltage by potentiometer,

I = Ammeter reading, |

E = EMF of the standard cell,

g_1 = length of potentiometer wire corresponding to the EMF of standard cell, and

l_2 = length of potentiometer wire for the potential difference V' or for current I' by potentiometer.

Procedure:

1. Insert the plug in the key K_2 so that current flows in the circuit connect the two ends of the standard one ohm resistance coil to the potentiometer by inserting the plug in the right hand gap (cb) of the two-way key K_3. The left hand gap (ac) of K_3 remains open.
2. Adjust the auxiliary circuit rheostat, so that ammeter and voltmeter read a suitable value, say 0.1. Note these reading of voltmeter V and ammeter I and simultaneously move the jockey J along the potentiometer wire and locate a balance or null point on it where the deflection in the galavanometer becomes zero. Note down the total balancing length l_2 of the potentiometer wire between the jockey and zero end point A. The multiplication of this balancing length(l_2) by the potential gradient K gives the exact value of potential bdifference (V') across the standard 1 ohm resistance coil. Since the resistance is exactly 1 ohm, hence the exact value of current I' flowing through it will be numerically equal to the potential difference across it. This gives the exact value of voltmeter V' and ammeter I' as measured by potentiometer corresponding to their respective value as measured directly by voltmeter and ammeter respectively.

3. Now adjust rheostat Rh_2 for new reading so that the voltmeter and ammeter read a suitable value say 0.2. Note these reading and simultaneously obtain balancing length l_2 as described above corresponding to new voltmeter and ammeter readings

Similarly, by changing the position of rheostat Rh_2 take several such observations in regular steps to (say) 0.1 (that is, 0.2,0.3,0.4,0.5......) till the entire range of voltmeter and ammeter is calibrated and note down all these readings.

Observations:

(1), Table A: For the calibration of Voltmeter and Ammeter

S N	Voltmeter reading V (volts) (a)	Ammeter reading I (amp.) (b)	Balancing length of the potentiometer wire from zero end A (l_2) in (cm)					$V'=I'=Kl_2$	V'-V (volts)	I'-I (amp.)
			No. of complete wires (n)	Distance of the null point from the intial end of the wire concern I(cm)		Total balancing length l_2				
1.										
2.										
3.										
4.										
5.										

Calculation: Potential gradient $K = \dfrac{}{}$ Volts/cm

$$V'= E = K \times \,.. = ... \text{ volt}$$

$$\text{and} \quad I' = E = K \times \,.. = ... \text{ amp}$$

Plotting of Graphs:

(i) Draw the graph for correction of voltmeter scale, between the error V'-V and voltmeter reading V. The slope of the graph would be zig zag as shown in Fig. 9.3.

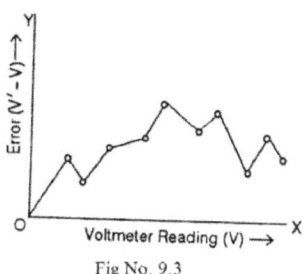

Fig No. 9.3

(ii) Now draw another graph for correction of ammeter scale, between the error I'-I and ammeter reading I. The slope of the graph will be zig zag as shown in Fig.9.4. The graph plotted between errors against the ammeter and voltmeter readings is the calibration curve.

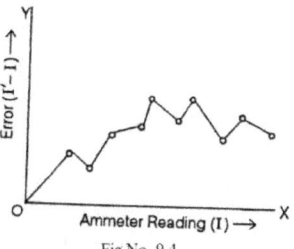

Fig No. 9.4

Result: The graph plotted between error against the ammeter and voltmeter reading is the calibration curve.

Precautions and Sources of Error:

1. The storage batteries should be fully charged.

2. The E.M.F. of the battery B_1 in the main circuit should be greater than that of the battery (B_2) in the auxiliary circuit.

3. The E.M.F. of the battery B_1 should be greater than the E.M.F. of standard cadmium cell or daniel cell.

4. The E.M.F. of the battery B_2 in the auxillary circuit should be greater than the range of voltmeter and ammeter which are to be calibrated.

5. Every battery and cell should be attached with a separate plug key and plug key should be closed at the time of their use.

6. The contact between the jockey and the potentiometer wire should be momentary.

7. The voltmeter and ammeter should be calibrated over its entire range.

Viva-Voce

Q.1. What are you doing?

Ans. Sir/Madam, I am calibrating a voltmeter and an ammeter by potentiometer.

Q.2. What is the principle of working of your potentiometer?

Ans. When a constant current is made to flow through the potentiometer wire, the fall of potential along it is directly proportional to the length of the wire. For this, area of cross-section of the wire should be uniform.

Q.3. What do you mean by potential gradient?

Ans. It is fall of potential per unit length of the potentiometer wire.

Q.4. Can you change the potential gradient?

Ans. Yes, it can be changed by changing the current flowing in the potentiometer wire by means of the rheostat (Rh_1) connected in the main circuit.

Q.5. What type of cell should be used in the main circuit?

Ans. The cell used in the main circuit must be of large and constant E.M.F. so that the potential difference across the potentiometer wire is more than the E.M.F. of the battery used in the auxilary circuit.

Q.6. Why do you always connect voltmeter in parallel?

Ans. A voltmeter is always connected in parallel because it has a very high resistance. Theoretically it has infinite resistance. Across the conductor a voltmeter does not draw any appreciable current and thus will not effect the current through the conductor or circuit and as such the potential difference across the conductor remains inaffected. Hence, it will measure correct potential difference across the conductor. If a voltmeter is connected in series it will decrease resistance in the circuit which is undesirable.

Q.7. What is the basic difference between the voltmeter and an ammeter?

Ans. A voltmeter has a very high resistance, whereas an ammeter has a very low resistance.

Q.8. What is the principle of working of a moving coil galvanometer?

Ans. It works on the principle that when a current carrying conductor is placed at right angle to a magnetic field, the conductor begins to move in a direction perpendicular to both the magnetic field and the direction of current.

Q.9. Why do you use copper connection wires?

Ans. Because, practically copper connection wire offer no resistance in the flow of current in an electric circuit, due to their very low specific resistance.

Q.10. What do you mean by standard cell?

Ans. A cell which has constant E.M.F. for a long time and low temperature coefficient is called standard cell.

Experiment No. 10

Object: To study the Hall effect and to determine Hall coefficient, carrier density and mobility of a given semiconductor (N-type) material using Hall effect set up.

Apparatus required: A rectangular slab, of about dimensions 7mmX2mmX0.2mm, specimen crystal, an electromagnet capable of producing magnetic field of the order of 10^3-10^4 gauss, a battery, two plug keys, milliammeter millivoltmeter, a voltmeter for measuring applied potential difference, rheostat, search oil, ballistic galvanometer to measure the magnetic field and connection wires.

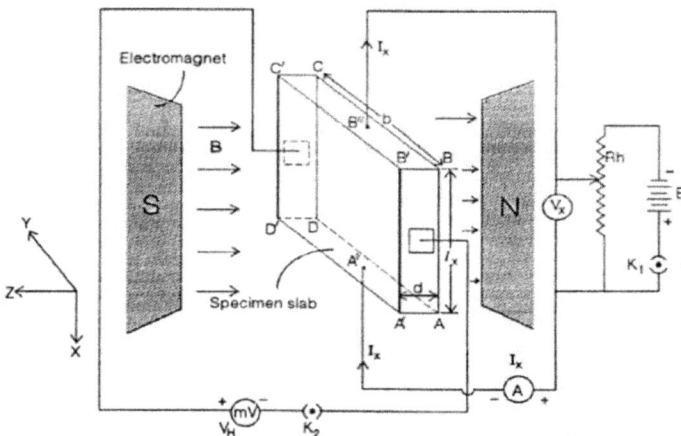

Fig No. 10.1

Description of Apparatus: The rectangular slab of semiconductor crystal (N-type) of dimensions about 7mm × 2mm × 0.2mm is placed between the pole pieces of a strong electromagnet capable of producing magnetic field of the order of 10^3- 10^4 gauss. A battery in series with a key K_1 is connected across the fixed terminals of the rheostat Rh used as a potential divider. The voltmeter V of the potential divider in series with an ammeter A is connected across the length of the specimen slab for measuring the applied voltage Vx and the current Ix passing in X- direction through it. To measure the Hall voltage V_H alongY-axis, a millivoltmeter in series with a plug key K_2 is connected as shown in Fig.10.1.

A stabilised source having variable voltage is used to fed current in the electromagnet. Either a calibrated fluxmeter is used to measure magnetic field or the deflection of a suitable ballistic galvanometer is calibrated to give the value of magnetic field.

Principle and Theory:

Principle: When a current carrying conductor is placed in a magnetic field, acting in a direction perpendicular to the direction of the current, a potential difference is developed across those faces of the conductor which are at right angle to both current and magnetic field. This phenomenon is called Hall effect and the induced potential difference is called Hall Voltage [Fig.10.2(a)].

Theory: The electric field E_x applied along X-direction exerts a force on the charge carriers due to which electrons acquire a drift velocity v_x in X-direction. The charge moving with drift velocity v_x experiences a magnetic force F_m perpendicular to both the velocity v_x and magnetic field induction B_z and is given by

$$\vec{v}\vec{x}=q(=\!x\!e\,() = (l_z\,r\!) = -\!\varphi = -ev_xB_z$$

This force acts along negative Y-axis due to which electrons on which it is exerted drift from positive Y-axis toward negative Y-axis. Thus, an induced electric field, called Hall electric field, E_H is established which causes a Hall potential difference V_H across the positive and negative Y-axes with positive Y-axis being at high potential difference [Fig.10.2(b)]. if the charge carriers are holes then the polarity of potential difference is reversed, that is, the surface towards the negative Y-axis would be at higher potential [Fig.10.2(c)] than the surface towards the positive Y-axis. Hall electric field so produced opposes the drifting of the electrons and very soon an equilibrium is reached due to the establishment of the condition of balance between the magnetic force $e(o \times h_l)$ on the electrons along negative Y-axis and the electric force eE_H along positive Y-axis and consequently drift velocity acquires a steady value. In this situation the net force acting on the electrons would be zero, that is,

$$\cdot(v = -(v \times B)$$

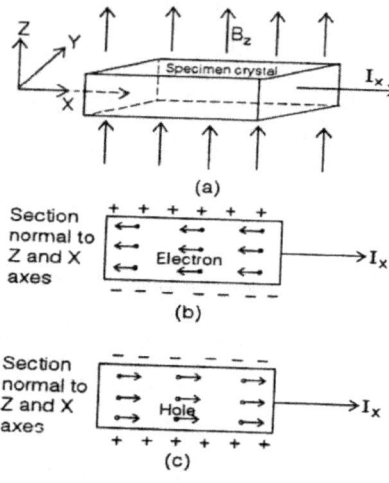

Fig No. 10.2

In terms of magnitude

$$E_z = v. B. \quad \ldots\ldots(1)$$

If b is the breadth of the specimen slab in Y-direction, then

$$V_H = bE_H$$

where, V_H is the Hall potential difference.

Now, $V_H = b \quad \ldots \quad \ldots\ldots(2)$ (from eq. (1))

If n is the number of charge carriers (electrons) per unit volume, then current density J_x is related with the drift velocity v_x of the electron as

$$V = -\frac{}{} \quad \ldots\ldots\ldots\ldots(3)$$

Therefore, $E^- = (-\frac{}{z})J.B. \quad \ldots\ldots(4)$

The quantity (1/ne) is called Hall coefficient for the material of the specimen slab and represented as

$$R_H = -1/ne$$

Negative sign appears due to the consideration of electrons as charge carriers.

From equation (4), we have

$$R_{\frac{}{R}} = \frac{}{} \dots\dots\dots\dots(5)$$

But $\quad \frac{}{\partial} = \dfrac{}{\text{cross section} \quad bd} = \frac{}{}$

$$R_{\frac{}{D}} = \frac{(\frac{V_H}{b})}{(\frac{I_x}{bd})B_z} = \frac{V_H}{} \frac{d}{B_z}\text{ohm-m}^3/\text{weber} \ \dots.(6)$$

where, B_z is in weber/m^2 and d in meter.

The ratio $\frac{}{}$ is determined graphically.

Formula used:

(1) Hall coefficient: $R_H = \frac{}{} . \frac{}{}$ ohm-m^3/weber.

If μ is the permeability of the meduim of the slab, then actual magnetic field within the rectangular slab is $B_z = \mu B$

(2) Number of charge carriers per unit volume n in the semiconductor slab is calculated as $n = \frac{}{}$

where, e is the charge on the electron or hole and equal to 1.6×10^{-19} colomb.

(3) Hall angle $\emptyset = \frac{}{}$ radian.

where, V_X is the potential difference applied across the specimen length.

(4) Mobility of charge carriers $m_\mu = \emptyset\,/B_z$ rad-m^2/weber

where, B_z is determined by a gaussmeter or ballistic galvanometer or fluxmeter.

(5) The electrical conductivity of the specimen slab α is calculated as

$$^\mu_r = \underline{\quad}^\mu\text{mho}\qquad.$$

Procedure:

1. Connect the Hall crystal to constant current power supply in their respective sockets.
2. Switch ON the power supply and adjust the current I. (say few mA)

3. There may be some voltage in the mV meter even outside the magnetic field. This is due to imperfect alignment of four contacts Ge crystal and is generally known as 'zero field potential'. In case its value is comparable to Hall voltage it should be adjusted to a minimum possible (for Ge crystal only). In all cases, this error should be subtracted from the Hall voltage reading.

4. Now place the probe in the magnetic field and switch on the electromagnets power supply and adjust the current to any desired value. Rotate the Ge crystal probe till it become perpendicular to magnetic field. Hall voltage V_H will be maximum in this adjustment.

5. Measure Hall voltage for both the directions of current and magnetic field.

6. Change the value of I_X in steps and note corresponding value of I_X and V_H. Take readings then plot a graph in V_H and I_X values. It will be straight line whose slop will be given by $\underline{\quad}$.

7. Measure the magnetic field B with a gauss meter.

Observations:

Length of the specimen slab along X-axis I_x = ... m
Breadth of the specimen slab along Y-axis b = ...m
Thickness of the specimen slab along Z-axis d = ...m
Permeability of the medium of the specimen μ =
Number of turns correspond to one division of the fluxmeter scale, n =......
Total number of turns in the search coil N = ...
Mean area of the search coil A = ...m^2

(1) Table A: For the Measurement of Magnetic Field B by Fluxmeter

S.No.	Current in the ammeter I (amp.)	Deflection in the fluxmeter	Magnetic field B =—
1.			
2.			
3.			
4.			

(2) Table B: For Measurement of Hall volatage V_H

S.No.	Magnetic field B measured with fluxmeter = ...weber/m^2 Actual magnetic field, B_Z = μB = ...weber/m^2		
	Current I_X in the specimen (amp.)	Applied potential difference V_X (volt)	Half voltage developed $V_H = V_Y$(volt)
1.			
2.			
3.			
4.			
5.			

Calculations: A graph is plotted by taking 9 along X-axis and the corresponding Hall voltage V_H along Y-axis which will be a straight line (Fig.10.3).

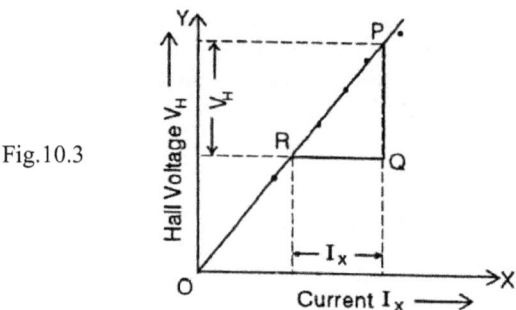

Fig.10.3

The slope of the curve = PQ/RQ = –– = ⋯

The Hall coefficient is $\dfrac{r_1}{1}$, $\dfrac{}{} = \dfrac{V_H}{I_x}$. ⟂ = ⋯ ohm – m³/weber

The number of charge carriefs per unit volume is n = $\dfrac{}{1.6 >}$ = $\dfrac{}{·19 \times_1}$ = ⋯

Hall angle, $\emptyset$ = –. ·. = …….. rad.

Mobility of charge carriers m_μ = $\dfrac{\emptyset}{}$ = ⋯ ⋯ $weler^{-1}$

The electrical conductivity of the specimen slab $\dfrac{u}{}$ = ––

Result:

1. Hall coefficient R_H = …. Ohm-m³/weber

2. Number of charge carriers per unit volume n = ……

3. Hall angle $\emptyset$ = ... Rads

4. Mobility of charge carries m_μ = ...rad-m^2/weber.

5. Electrical conductivity of the specimen slab .n = ...mho/m

Precautions and Sources of Error:

1. The distance between the pole pieces of the electromagnet should not be changed during the whole experiment.

2. Magnetic field should be kept constant for one set of observations.

3. Current passing through the experiment slab of the semiconductor should be strictly within the permissible limit.

4. Hall voltage should be measured very carefully and accurately either by a millivoltmeter or by a potentiometer.

VIVA-VOCE

Q.1. What are you doing?

Ans. Sir/Madam, I am studying Hall effect in a semiconductor.

Q.2.What is Hall effect?

Ans. When a current carrying conductor is placed in a transverse magnetic field a potential difference is developed across the conductor in the direction perpendicular to both current and magnetic field. This phenomenon is called Hall effect.

Q.3. On what factor the sign of Hall potential difference depends?

Ans. The sign of Hall potential difference depends upon the nature of charge carriers. It decides whether a specimen semiconductor is of N-type or P-type.

Q.4. What is Hall coefficient?

Ans. Hall coefficient, R_H is numerically equal to the Hall electric field E_H induced in a specimen crystal by unit current density when it is placed transversly in a magnetic field of a 1 weber/m^2.

Q.5. What is unit of Hall coefficient?

Ans. Unit of Hall coefficient R_H is ohm-m^3/weber or m^3/coulomb.

Q.6. What are the importance of hall effect?

Ans. Hall effect has following importance:

1. The sign of the Hall potential difference developed determines the nature of the charge carriers.
2. The mobility of charge carriers is measured directly.
3. It can be used to decide whether the specimen is a metal, semiconductor or an insulator.
4. With the help of Hall coefficient R_H the number of charge carriers per unit volume can be determined.
5. With the help of Hall potential difference one can calculate the magnetic field.

Q.7. What is Hall angle?

Ans. When a transverse magnetic field is applied to a current carrying conductor, then the charge carriers come under the influence of simultaneous applied electric field E_x and induced Hall electric field E_H at right angles to each other. In this situation the angle made by the drift velocity of the charge carrier with the X-direction is called Hall angle and represented as

$$\Phi = -\underline{\quad}$$

Q.8. What do you mean by mobility of a charge carrieer?

Ans: The mobility of charger carrier is defined as the drift velocity acquired by a unit applied electric field. It may also be defined as the ratio of average drift velocity of the charge carriers to the applied electric field.

Q9. What is the unit of mobility of charge carriers?

Ans: The unit of mobility is m^2/(volt-sec) or radian-m^2-weber^{-1}.

Q.10. Which type of the charge carrier has greater mobility?

Ans. In the semiconductor, the electron mobility is greater than hole mobility.

Experiment No. 11

Object: To determine the energy band gap of given extrinsic semi-conductor material by thermal variation using four probe method.

Apparatus required: Four probe arrangement, four probe set up, oven with supply, germanium or silicon crystal chip with non-conducting base and a sensitive thermometer.

Description of Apparatus:

Four Probe Arrangement with Oven: In four probe method four spring type contacts are used to avoid contact resistance. In fact soldered probes contact or direct soldering to the body of the sample effects the sample properties by heating effect and contamination. Four electrodes probe arrangement consists of four spring type collinear equally spaced probes coated with zinc at the tips (Fig.11.1). The probes are mounted in a teflon bush for good electrical insulation. The outer probes (1,4) are used to pass current I through the specimen of known conductivity sample S^1. The constant current source I used in the experiment is specially designed for this method to provide 100% protection from crystal burn due to excess current. In this source the value of current can be changed by the potentiometer included for that purpose. The current I is usually low, of the order of milliampere and is measured by a milliammeter of 0-10 mA range. The inner pair probes (2,3) measure voltage V by a digital electronic millivoltmeter specially designed for this purpose. However, any pair of electrodes may be used to pass current I while the remaining pair may be used to measure voltage V. The whole arrangement is

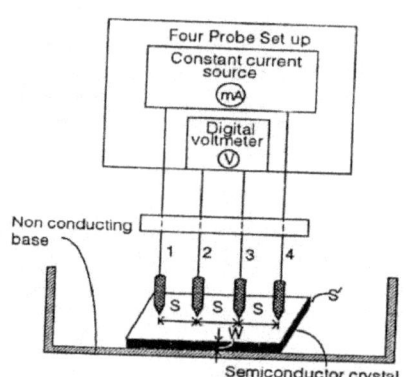

Fig. No. 11.1

mounted on a suitable stand and leads are provided for current and voltage measurements (usally green leads are provided for current and orange leads for voltage measurement). At the top of the four probe arrangement stand a hole is provided for inserting a thermometer to measure the oven temperature. A good quality thermometer of generally 0-200°C range is used to measure the oven temperature. Variation of temperatures of crystal from room temperature to about 175°C is studied in a small oven the supply for which is built-in inside the set up. The supply has three different output voltages for the oven to change the rate of heating.

Assumptions for Four Probe Technique: The four probe technique is based on the following assumption:

1. The surface of the sample on which the four electrodes probe rest is flat, uniform, having no leakage, adequately large and crystal should be of big size.
2. The spring type contacts between the probes are point contacts and lie in a straight line.
3. The resistivity of the crystal is considered to be uniform in the area of measurement.
4. The bottom surface of the sample chip is non-conducting.
5. The contacts between the probes and the surface are point contacts.
6. The thickness of the sample in chip shape should be less than the half the distance between the probes or probe spacing(S).
7. If there is minority carrier injection into semi-conductors by current carrying electrode probes most of the carrier recombine near the electrode probes most of the carrier recombine near the electrode probes so that there effect on the conductivity become negligible.

Theory and Formula used: If the current I is passed through the outer probes and V be the potential difference measured across the inner probes, then the resistivity the sample is given by

$$\rho_0 = \frac{V}{I} \times 2\pi S$$

where, S is the distance between the two successive probes.

If probe spacing is 0.159 cm, then

$2 \quad = 1 \quad$ and $\quad \rho_0 = -$

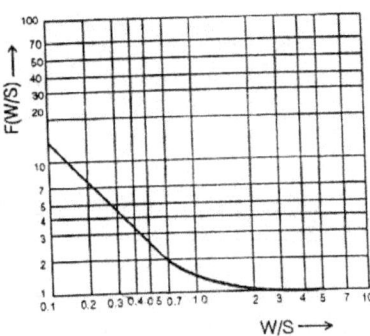

Fig. No. 11.2

For a thin slice non-conducting bottom surface, the restivity may be expressed as

$$P = \dfrac{}{(-)} \cdot$$

Where W is the thickness of the semi conducting material and S the probe spacing. The value of function is $(-)$ is obtained either from Table A or from the graph is shown in Fig. 11.2. in table A, the value of F(W/S) for different combination of W/S is given. If any W/S value is not found in the table, then the desired value of F(W/S) corresponding to any value of (W/S) can be obtained from the graph shown in Fig. 11.2.

Table A		
S.No.	W/s	F(W/s)
1.	0.100	13.863
2.	0.141	9.704
3.	0.200	6.931
4.	0.333	4.159
5.	0.500	2.780
6.	1.000	1.504
7.	1.414	1.223
8.	2.000	0.094
9.	3.333	1.0228
10.	5.000	1.0070
11.	10.000	1.00045

We know that forbidden energy band gap E_g of semi-conductor is related with the resistivity as

$$Log_e\rho = \text{———}$$

or $\quad E_g = 2k\,\dfrac{'\underline{\varsigma_{10}}\,\rho}{/}\quad -$

Where k is the Boltzmann's constant equal to 8.6×10^{-5} eV/deg and

$$Log_e\rho = 2.3026\; Log\;\;\rho.$$

Thus the slope of the graph between $log_{10}\rho$ and $10^3/T$ gives

$$\dfrac{\text{bs}}{2/.}\;\; - = \dfrac{-}{\text{-s}}\;\; \text{or } E_g = 2k \times (\dfrac{'\underline{\varsigma_{10}}\,\rho}{'/}\;\; -).$$

Procedure:

1. First of all the four probe arrangement is taken out of the oven and keep it on a plane surface. Now the sample crystal is put on the circular base plate of the arrangement in such a way that its non-conducting surface lies towards the plate side and four probe are in the middle of the sample crystal. To make the proper contact of the probes with the sample crystal apply slight pressure on their pipes and screws are tightened.
2. Now check the continuity between the sample crystal and four probes with the help of multimeter. For proper contacts the resistance between the outer probes leads and inner probes leads should be nearly equal and of the order of $1K\Omega$ to $3.5K\Omega$. If the contacts are not proper, tighten the screws provided on the top of the base stand till the probes touch the sample crystal.
3. Now keep the four probe arrangement on the oven properly and again check the continuity between the probes as described above.
4. Now insert the thermometer in the hole provided in four probe arrangement. Connect the outer probe leads to the output terminals of the constant current source and the inner probes leads to the input terminals of the digital

millivoltmeter of the four probe set up. Now connect the set up to the A.C mains and put the switch to 'ON' position.

5. In the four probes set up, put the selector switch on current position and adjust current to zero value with the help of current adjustment knob. Now change the selector switch in the voltage position and adjust zero value in digital voltmeter by shorting the voltage terminals.

6. Again change the selector switch of set up towards current position and apply some current say about 5mA by means of current adjustment knob and keep this current value constant by undisturbing current adjustment knob for one set of observation. Now change the selector switch again towards voltage position.

7. Connect the oven with the oven supply and adjust its switch to lower position.

8. As the oven is switched on, its temperature rises. in thermometer. Note voltage in digital voltmeter and corresponding temperature in thermometer. When the temperature rises above 45^0C put the oven supply switch in medium position. Note voltage in digital voltmeter and corresponding temperature in thermometer. Again above 90^0C, the oven supply switch is adjusted to higher position.

9. The voltage for different values of temperatures keeping current constant are recorded upto the temperature of about 160^0C.

10. Repeat the above procedure for different values of current.

Observations:

Thickness of the crystal,   W = ...cm

The distance between the probes, S = ...cm

Value of current,   I = ...Ma

Table A:

S.No.	Temperature of the sample		Corresponding voltage V (volts)	Resistivity $\rho_0 = \frac{1}{\times 2}$ (ohm-cm)	Corrected resistivity P = — - (Ohm-cm)	... >	$\frac{()}{10^3}$ ×
	In (^{0}C) T(^{0}C)	In Kelvin T(K)					
1.	20^{0}C	293					
2.	25^{0}C	298					
3.	30^{0}C	303					
4.	35^{0}C	338					
5.	40^{0}C	343					
6.							
	160^{0}C	437					

Calculations:

First of all the resistivity ρ_0: corresponding to temperature in Kelvin is found by using the relation

$$\rho_0 = \frac{V}{I} \times 2\pi S$$

The values of I and S are taken from observations.

Now corrected restivity ρ for probs on a thin slice with a non-conducting bottom surface and for the thickness W of the crystal is determined by using the relation

$$\rho = \frac{- -}{(-)}$$

The value of function, F(W/S) may be obtained from the Table A or graph (Fig.11.3) for appropriate value of W/S. Hence, calculate ρ corresponding to different values of temperatures.

Fig. 11.3

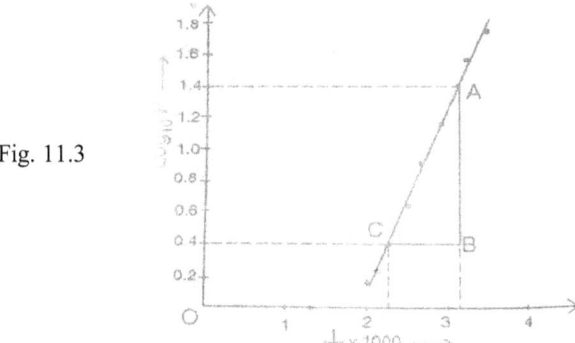

Now a graph between $(1/T) \times 10^3$ and $\log_{10}\rho$ is plotted. The graph will be a straight line (Fig.11.3). The slope of the curve will be

$$\frac{L}{\left(\frac{1}{=}\right.} = \; ; \; - \; - \; -$$

Now, forbidden energy band gap of semi conductor is

$$E_g = 0.396 \; (\text{—}) \; eV$$

Percentage Error: The percentage error in the experimental value is calculated by the following formula

$$\text{Percentage error} = \frac{\text{'d value–Calculated value}}{\text{'d value}} \times 100\% = \ldots..\%$$

Result:

Forbidden energy band gap for semi conductor is $E_g = \ldots..eV$.

Standard Result:

Standard value of forbidden energy band gap of semi conductor (Germanium) is

$E_g = 0.67$ eV.

Precautions and Sources of Error:

1. The resistivity of the sample crystal should be uniform in the area of the measurement.
2. The sample surface should be uniform and having no leakage.
3. The sample should be placed with non-conducting surface towards bottom.
4. There should be proper contact between the probes and sample crystal surface.
5. Current should be constant for one set of observation.
6. The current through the sample should not be large enough to cause heating.

Viva-Voce

Q.1. What are you doing?

Ans. Sir/Madam, I am determining the energy band gap of extrinsic semi-conductor material by thermal variation using four probe method.

Q.2. What do you mean by energy band gap?

Ans. A measure of energy gap between valance band and conduction band is called energy band gap

Q.3. What is n-type semi conductor?

Ans. By adding pentavalent impurity to pure silicon or germanium, number of free electrons increase. Such an impure semi conductor having excess of electrons as charge carrier is called n-type semi-conductor.

Q.4. What is p-type semi conductor?

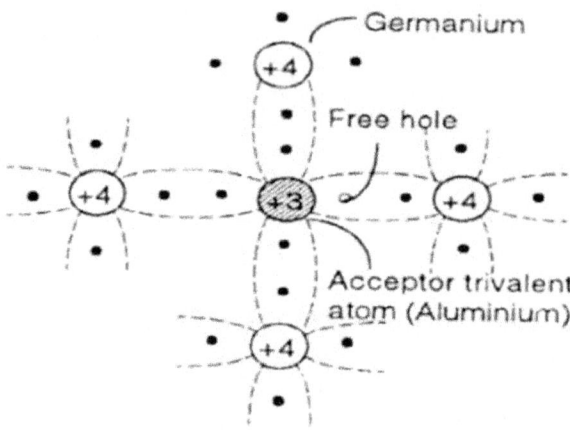

Fig.11.4

Ans. A germanium or silicon crystal with trivalent impurity is called p-type semi-conductor because the charge carriers are positive hole (Fig. 11.4).

Q.5. What is n-type semi-conductor?

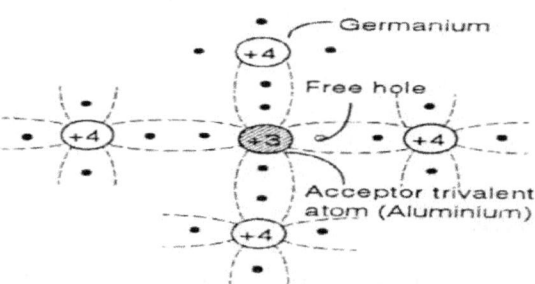

Fig. 11.5

Ans. An n-type semi-conductor is formed when an impurity atom with 5 valence electrons is introduced in a pure germanium. It replaces a germanium atom and four of the five valence electrons of the impurity atom form covalent bond with one each valence electron of four germanium atom and the fifth valence electron becomes free. This free electron acts as a charge carrier. Hence, by adding pentavalent impurity to pure germanium the number of free electrons increases. Such an impure germanium crystal having excess of electrons as charge carrier is called n-type semi-conductor (Fig. 11.5).

Q.6. What are the intrinsic and extrinsic semi-conductors?

Ans. In intrinsic semi-conductor, electrons and holes are solely created by thermal excitation and numbers of free electrons and holes are equal.

It has small electrical conductivity. Pure germanium and silicon are the example of intrinsic semi-conductors.

When a small amount of pentavalent or trivalent is added to the pure semi-conductor, the conductivity of the semi-conductor is significantly increased. These semi-conductor are called extrinsic semi-conductors.

Q.7. Why is the four probe method is better than other methods for measuring the resistivity?

Ans. In this method, the current carrying contacts inject minority carriers which changes the resistance of the material.

Q.8. What is band gap in a good conductor?

Ans. In a good conductor there is no band gap as the conduction and valence bands overlap in good conductors.

Q.9. What do you mean by forward and reverse biasing of a junction diode?

Ans. When a battery is connected to the diode with p-section connected to the positive pole and n-section to the negative pole,then the junction diode is said to be forward biased. On the other hand, when a battery is connected to junction diode with p-section connected to negative'pole and n-section to the positive pole, the junction is said to be reversed biased.

Q.10. What happens just after the formation of junction?

Ans. Just after the formation of junction, the free electrons which are the majority charge carriers in the n-region, and holes that are the charge carriers in the p-region, tend to diffuse to their oppostie sides, that is, free electron to p-side and holes to n-side.

Experiment No. 12

Object: To determine the electro-chemical equivalent (E.C.E.) of copper and redusction factor of Helmholtz Galvanometer.

Apparatus Required: Balance, copper voltmeter with four copper plate, a storage battery, Helmholtz galvanometer, commutator, A plug key, A rheostat, stop watch, connection wire, accumulator.

Description of Apparatus and Circuit:A copper voltameter (Fig. 12.1) consists of a cylindrical glass vessel filled nearly (2/3)rd by a solution of copper sulphate

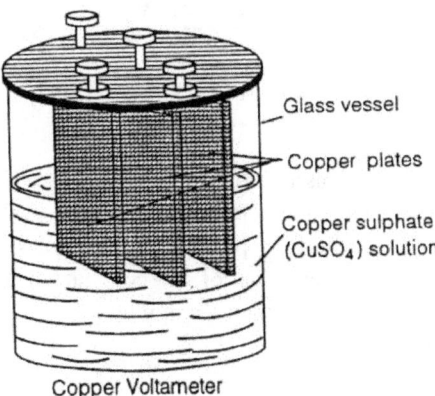

Fig No. 12.1

($CuSO_4$) with a few drops of sulphuric acid. The vessel is covered with a circular ebonite lid from the top of which anode and cathode copper plates are suspended. The anode is a pair of two identical parallel copper plates held at a small distance apart and joined together by a metallic strip with the help and of binding terminals provided at their upper ends. The cathode is also a copper plate suspended symmetrically between the two anode plates through the central hole in the lid by means of a binding terminal provided at their upper end. When an electric current is passed through the copper sulphate solution, copper deposited on the cathode and dissolves from the anode into the solution keeping the concentration of the solution constant.

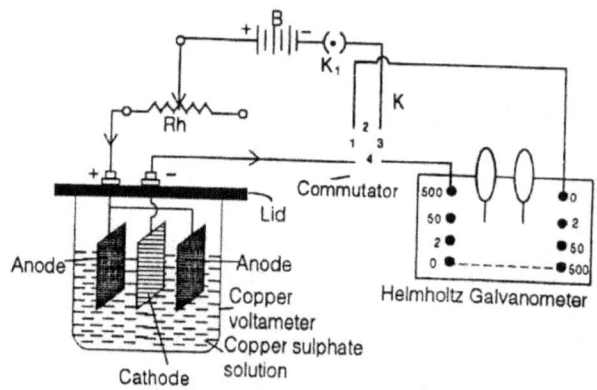

Fig. No. 12.2

The scheme of connections is shown in a circuit in a diagram depicted in Fig.12.2. The two terminals of a Helmholtz galvanometer (or a Tangent galvanometer) are connected to two diagonally opposite fixed terminals of the commutator K. The remaining two movable terminals of the commutator are connected to the copper voltameter in series with a rheostat Rh, a plug key K1 ε and a storage battery B as shown in the Fig. 12.2.

Formula used: If the coil of the Helmholtz galvanometer coil be set in the magnetic meridian and a current of strength I ampere is allowed to pass through it, then

$$i = \frac{\sqrt{}}{\ } \tan\theta$$

where,   r = radius of the coil,

n = number of turns used in the coil of Helmholtz galvanometer, and

H = Earth's magnetic field's horizontal component.

If the same current is allowed to flow through a copper voltmeter connected in series with Helmholtz galvanometer, then from Faraday's law of eletrolysis

$$m = Zit$$

Where   m = mass of the copper deposited on cathode plate,
         t = time in seconds for which the current flow,
         i = strength of the current, and
         Z = electro-chemcal equivalent of copper

Therefore, $z = \dfrac{}{\sqrt{!}} = \dfrac{}{I\ t_{ε_\text{M}}\text{Ext} \quad \times}$

the reduction factor of Helmholtz galvanometer

$$k = \dfrac{}{\rule{2cm}{0.4pt}} \ .$$

Procedure:

1. Clean the cathode plate thoroughly with sand paper and weight it with the help of chemical balance.
2. Place the coils in magnetic meridian. Rotate the compass box to make the pointer read zero.
3. Using copper test plate as cathode, allow a current to flow in the circuit and read the deflection. Now reverse the current wth the help of commutator and again read the deflection. If the two deflection are the same then the coils are in the magnetic meridian otherwise rotate slightly the coils till the two deflection are the same. The pointer should read zero when no current is passed.
4. Using rheostat Rh adjust the deflection say within 45° to 50°.
5. Switch off the current and remove test plate. Now place the previously weighted plate to act cathode.
6. Switch on the current and immediately start the stop watch. Note down the deflection after a regular interval of five minutes and keep it constant with the help of rheostat. After 15 and 20 minutes reverse the current and note the deflection. At the end of other half of time, switch of the current and note down the reading of stop watch.

7. Remove the copper plate from voltammeter. Immerse the plate in the water and then press it between sheets of filter paper to soak the water. Dry it with the help of cold air blower and weight it with chemical balance.
8. Measure the diameter of coil.

Observations:

Number of turns in each coil (n) = ...

Radius of the coil (r) = ...cm

Value of the field (H) = ...oersted

(1)Table: For the determination of the mass deposited (m) and time taken (t)

S.No.	Quantities measured	Amount	Calculated quantities from observation
1.	Mass of the copper plate before deposition of copper	........gm	Mass of copper deposited m =......gm
2.	Mass of the copper plate after deposition of copper	........gm	
3.	Initial reading of stop watch	........sec	Total time t taken = ...sec
4.	Final reading of stop watch	........sec	

(2)Table B: Table for determination of

Time (in minutes)	Deflection of pointer for direct current		Deflection of pointer for reverse current		Mean	Tan
	Left end	Right end	Left end	Right end		
0						
5						
10						
15						
20						

Calculations: $Z = \dfrac{1}{Ita\sqrt{\theta}\times t \qquad \times} = $ .....gms/coulomb

The reduction factor of Helmholtz galvanometer

$$k = \underline{\quad\quad} = \text{.....amp.}$$

Result: The electro chemical equivalent of copper = ...gms/coulomb
The reduction factor of Helmholtz galvanometer = ...amp.

Standard Result: The standard value ofE.C.E. of copper =0.0003295 gms/coulomb.

Percentage Error: The percentage error in the experimental value is calculated by the following formula

Percentage error $= \dfrac{\text{·d value}-\text{Calculated value}}{\text{·d value}} \times 100\% = $ .....%

Precautions and Sources of Error:

1. The connection wire should be thick and their ends should be properly cleaned with sand paper before use. All the connections to the binding terminals should be tightly made.
2. The experimental copper plate should be thoroughly cleaned with sand paper and running water and than make free from grease otherwise the copper deposited on it will not adhere ptoperly.
3. Keep all the current carrying conductor and magnetic material far away from the Helmholtz galvanometer otherwise galvanometer gives incorrect observations.
4. The current in the coil remain constant throughout the experiment and it should be so adjusted that the deflection in the needle of the galvanometer is about 45°.
5. The middle plate of the copper voltmeter should be used as cathode plate and should be connected to the negative terminal of the storage battery B.
6. No time should be lost in reversing the direction of current.
7. A few drops of sulphuric acid should be added to copper sulphate solution.

Viva- Voce

Q.1. What are you doing?

Ans. Sir/Madam, I am determiming the electro-chemical equivalent (E.C.E.) of copper using Helmholtz galvanometer.

Q.2. What do you mean by E.C.E. of an element?

Ans. The electro-chemical equivalent of an element is defined as the mass of its ions liberated at an electrode, when a charge of one coulomb is passed through its electrolyte or when one ampere of current is passed for one second through its electrolyte.

Q.3. What is an electrolyte?

Ans. The liquid or salt solution which undergo decomposition on passing current through it, is called electrolyte.

Q.4. What do you mean electrolysis?

Ans. The process of decomposition of a liquid or a salt solution by passing a current through it, is called electrolysis. During the process of electrolysis the positive ions are liberated at cathode and the negative ions are liberated at anode.

Q.5. Why do we add a few drops of sulphuric acid to copper sulphate solution?

Ans. By the addition of sulphuric acid the conductivity of the copper sulphate solution increases. It produces additional ions in the solution.

Q.6. What is chemical equivalent?

Ans. Chemical equivalent of an element is its that mass which combines or displaces one part by weight of H_2 or 35.5 parts by weight of Cl_2 etc. It is equal to the ratio of atomic weight to the valency of the element.

Q.7. Why do you make the deflection nearly 45^0?

Ans. It is because the value of tanθ does not change much near the angle 45^0. Therefore, a small error in the measurement of deflection does not affect the result appreciably.

Q.8. During the process of electrolysis, what happened to the concentration of the solution ?

Ans. During the process of electrolysis the concentration of solution remains constant because when electric current is passed through the solution, the copper deposited on the cathode and dissolves from the anode into the solution.

Q.9. Give the construction of tangent galvanometer?

Ans. It is based on the principle of tangent law. It consists of three circular coils of 2, 50 and 500 turns of insulated copper wire wound on a common non-magnetic circular

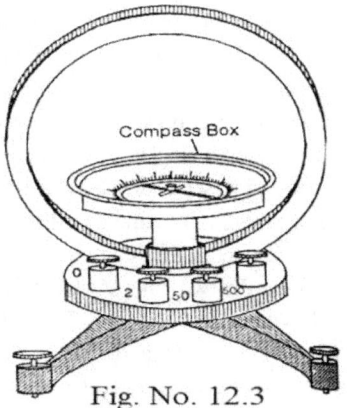

Fig. No. 12.3

frame held vertically on circular base (Fig. 12.3). The circular base can be rotated about a vertical axis and can also be levelled with the help of three levelling screws provided at the base of the galvanometer. A compass box, at the centre of which a small magnetic needle and long aluminium pointer is pivoted, is placed at the centre of the coil.

Q.10. Give the construction of Helmholtz galvanometer?

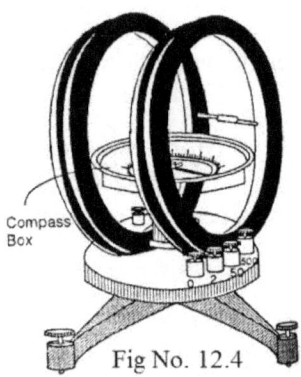

Fig No. 12.4

Ans. The working of a Helmholtz galvanometer is based upon tangent law. It consist of two similar coils of same radius and of the same number of turns (Fig.12.4). Each circular coil is a combination of three coils of 2, 50, 500 turns of insulated copper wire. These coils are mounted coaxially on a common circular base and wound on non-magnetic frames. The separation between the coils is equal to the radius of the either coil. The coils are connected in series and their windings are such that the same current passes through each of them in the same direction. A compass box, at the centre of which a small magnetic needle and long aluminium pointer is pivoted, is fitted at the central point between the two coils. At the bottom of the base three levelling screw are provided to level the apparatus. The galvanometer is provided with four binding terminals marked 0, 2, 50 and 500. Internally two turn coil is connected between the first and second terminal, 50 turn coil is connected between the second and the third terminal and 500 turn coils is connected between the third and fourth terminal respectively.

Experiment No. 13

Object: To Draw hysteresis curve (B-H curve) of a given sample of a ferromagnetic material in the form of thin iron wires (or cycle spokes) on a cathode ray oscilloscope (C.R.O.) ,using a solenoid and to determine retentivity, coercitivity and hystersis loss from it.

Apparatus required: Cathode Ray Oscilloscope (C.R.O.), Cycle Spokes of Iron or Rods of 40-50 thin iron wires, Solenlid, Step down Transformer, Rheostat (10   ), two condensers (of capacity 1μF and 2 μF), A.C. Ammeter, Two One Ohm Resistor, One 47 Ko Carbon Resistor, Two Carbon Potentiometers of 20 K 5 and 5K Resistance etc.

Theory and Formula used: When an A.C. current passes through the prism of the solenoid in which specimen rods are kept, then the magnetising field produced by solenoid's primary is given by

$$H=\text{---}\text{---}\cos\frac{\cdots v}{4}=\overline{\;\;ns)(\;\;\cdots\;.+\;}^{2}\cos$$

where n is the number of turns per unit length in the primary, I_{rms} is the root mean square value of current directly measured in A.C. ammeter. Its maximum value is

The voltage developed across 1 ohm resistor R1 is proportional to H and applied to x-x plate of C.R.O.

The induced voltage (E.M.F.) across the secondary coil is measured of dB/dt. Hence to get magnetic flux density B through it an integrating circuit (R-C) is used. In the circuit arrangement 47 Kto resistor and 2 μF capacitor constitute this integrated circuit. The induced e.m.f produced across the secondary is

$$e=-\text{---}\frac{\cdot}{7.4}(B)=\text{---}\frac{\cdot}{7}(BAN) \qquad \text{or} \qquad e=\text{-AN}\text{---}$$

It's magnitude is e =AN —   ....(1)

where A is the area of cross-section of the solenoid, N the number of turns in the secondary.

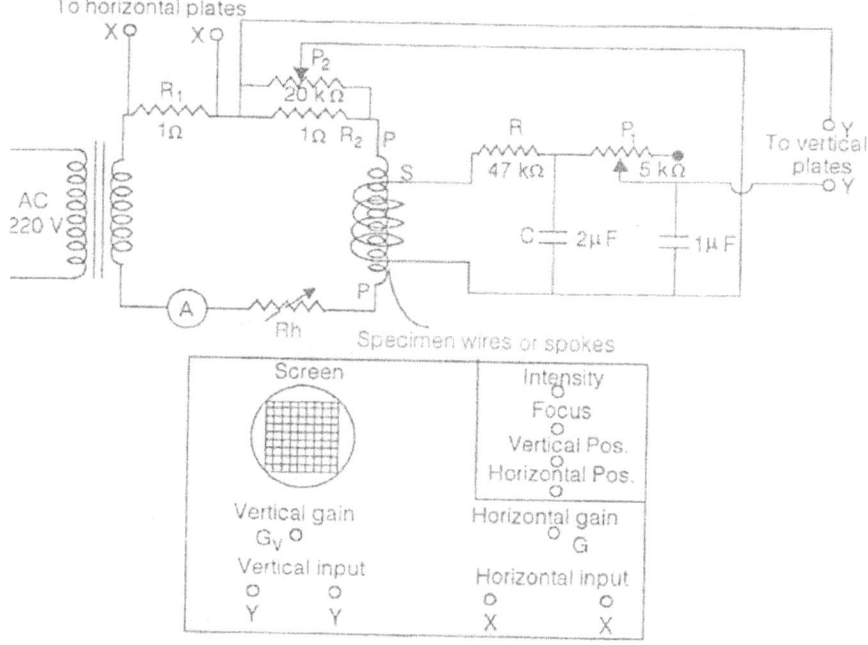

Fig. 13.1

The instantaneous current i flowing in the circuit is

$$i = \frac{\epsilon}{(-)} \approx -$$

The voltage drop across the capacitor C or applied to the pair of Y-Y plate of C.R.O is

$$V = \frac{\cdot}{\cdot} = -\iint \frac{\cdot}{n} = -\int - \, dt$$

From equation (1), we have

$$V = \frac{a}{d} \int t \quad —dt$$

$$\text{Or } V \propto —$$

The voltage proportional to B when applied to Y-Y plates a hysteresis loop is obtained on C.R.O.

Procedure:

Calibration of X-and Y- axes in terms of H and B

(A) Display of B.H. curve on the screen

1. After making necessary connections in accordance to the circuit diagram shown in Fig 13.1 A.C. supply is switched on and rheostat Rh is adjusted to get maximum current in the primary of the solenoid.

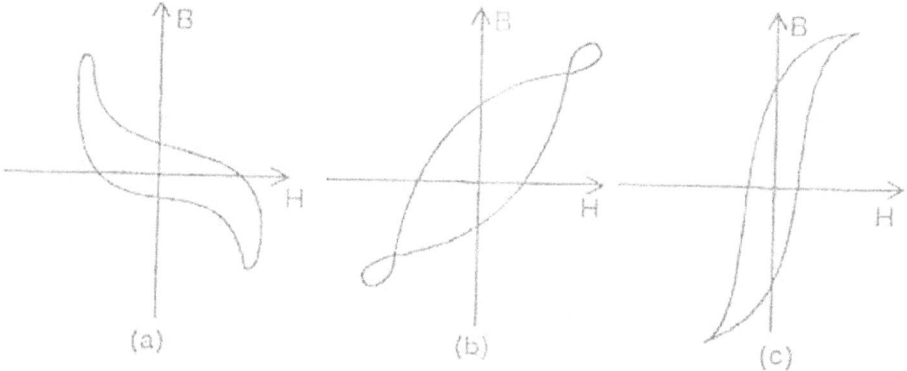

Fig. 13.2

2. Adjust the vertical and horizontal gains of the C.R.O with the help of gain control switches, G_h and G_v to get the proper shape and size of the wave form on the screen. Usually the wave form obtained on screen is not of the desired shape. To get the correct shape following adjustment are made:

(i) If the shape of the curve obtained on the screen is as shown in Fig. 13.2(a), then interchange secondary leads (Y-Y) to C.R.O to bring the wave form in the proper quadrant.

(ii) If the shape of the curve is in accordance to the Fig 13.2 (b), then the potentiometer P_1(adjusted such that the curve becomes free from loops or flatness at the tips of the pattern.

(iii) If the curve obtained on the screen is similar to the Fig 13.2 (c) but ends of the curve are not parallel to H axis, then adjust the potentiometer P_2(20 Ka) to make the ends horizontal. moreover, if an increase in the resistance of P_2 increase the slope at the

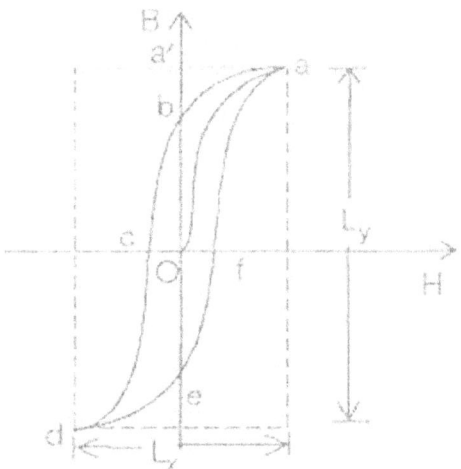

Fig. 13.3

ends instead of reducing, then interchange the connection to Y-plates. For these adjustments X and Y amplifier of the C.R.O may be varied freely to get correct shape (Fig 13.3).

(B) Tracing of B-H curve :

1. When the Shape of the B-H curve on C.R.O. screen becomes proper and correct (Fig 13.3), a Tracing paper is put on the screen. Now, vertical gain is reduced to zero with the help of gain control switch, G_v by keeping horizontal gain Maximum so that a straight line which indicates the H-axis is obtained on the Tracing paper.

2. Similarly, horizontal gain is reduced to zero with the help of horizontal gain control switch G_h by keeping vertical gain Maximum so that a straight line which indicates B-axis is obtained on the Tracing paper. Now adjust the horizontal and vertical gain control switches to their previous positions to get a B-H curve of proper shape. Now trace this B-H curve on the paper (Fig.13.3) for the calculation of various parameters.

(C) Calibration of X-axis in terms of H

Measure the maximum deflection of spot along X-axis on tracing paper, let it be L_x hence ,the calibration constant C_H for H-axis is

$$C_H = \text{————} \text{ oersted/cm}$$

(D) Calibration of Y-axis in terms of B

1. To calibrate Y-axis in terms of B, reduce horizontal gain to zero with the help of gain control switch G_h and also adjust potentiometer (20 K) P to the zero position so that no resistance acts in the secondary coil circuit. By doing so a vertical line equal to the cur left. Measure this as L_y.

2. Now take the specimen (cycle spokes or iron wire rods) slowly out of the primary of the solenoid till the length of vertical line reduces to its half value, that is reduces to $L_y/2$.

3. Now keeping the position of the specimen rod unchanged increase the vertical gain with the help of G_v to restore back the original length of the vertical line, that is, L_y. this implies that vertical gain is now become double to its previous value.

4. Now the experimental rods or specimen is further taken out of the primary till the length of the vertical line further reduces to ·y, again kegain keeping the position of

specimen rod unchanged further increase the vertical gain (G_v) to restore back the original length L_y of the vertical line again. At this time the vertical gain becomes four times to its initial value.

5. Repeat the process once or twice more so that vertical gain G_v becomes 8-times or 16-times to its initial value provided specimen is not totally out of the solenoid's primary. Let this amplification factor be F=2,4,8,16.. depending upon the steps taken to increase it.

6. Now the specimen is taken out of the solenoid completely and measure the length of the vertical line on the screen. Let it be h. The length of the vertical line h in the absence of specimen in the solenoid with original amplification will be h/F.

The vertical deflection h/F corresponds to double the, maximum magnetic flux due to air core (Φair)max or in the absence of any specimen in the solenoid's primary is given as

$$2(\chi_{air})max=2H_{max}\ r^2$$

where r is the mean radius of the primary coil.

Therefore ,the calibration constant for flux measurement is

$$C= \text{---}a\text{---}\ maxwell/cm$$

If B is the magnetic induction in the iron wires (or spoken or specimen) and S their area of cross-section, then

$$\text{1.}S_{max}=B.S$$

Hence, the calibration constant for B is

$$C_B=\frac{16}{n} = \text{---}a\text{---}\ gauss/cm$$

where $H_{max}= \text{---}$Oersted.

Observations:

(a) Constants for calibration of X-axis in terms of H

Current in the primary of the solenoid (ammeter reading), I_{rms}=... amp

Number of turns on the primary of the solenoid N = ...

Length of the primary l = ... cm

No of turns per unit length of the primary, $n = N/l$ = ... turns/cms

(b) Constants for calibration of Y-axis in terms of B

Mean radius of the primary of the solenoid r =... cm

Mean radius of one specimen wire or spoke R = ... cm

Total number of iron wires (or spokes) used N_1 = ...

Total area of cross-section of all specimen wires or spokes $S = (=R^2)N_1$= ...cm^2

The vertical gain or amplification factor F = ...

Length of the vertical line on the screen with air as core h = ...cm

(b) From the trace of B-H curve

Horizontal length of the curve L_x=...cm

Vertical length of the curve L_y=...cm

Calculations:

1. Calibration constants

Maximum value of magnetic field.

$$H_{max}=\frac{ns\quad\sqrt{}}{...}\cdots =... \text{ oersted}$$

Calibration constant for X-axis in terms of H

$$C_H = \text{—. ~~} = \text{... oersted/cm}$$

Calibration constant for Y-axis in terms of B

$$C_B = \frac{\ddot{}F}{\text{—}} \text{... —} = \text{.... gauss/cm}$$

2. Magnetic constant from B-H curve

From fig. 15.6

$$Ob = \text{...cm}$$

$$\text{Retentivity} = Ob \times C_B = \text{... gauss}$$

$$O_1 = \text{...cm}$$

$$B_{max} = O_1 = \times C_B = \text{... gauss}$$

$$Oc = \text{... cm}$$

$$\text{Coercivity} = Oc \times C_H = \text{... Oersted}$$

3. Area of the B-H curve

Area enclosed by B-H curve on the screen $S' = \text{...cm}^2$

Hence, the hysteresis loss per cycle per unit volume = — (area of the B-H curve)

$$= \text{—}(S' \times C_H \times C_B)$$

$$= \text{... ergs per cycle per cm}^3$$

Result:

1. The hysteresis curve (B-H curve) for given ferromagnetic material (say iron) is shown on the attached tracing paper.
2. Retentivity of ferromagnetic material such as iron   = ... Gauss
3. Coercivity of ferromagnetic material   = ... Oersted

4. Energy loss per cycle per unit volume or Hysteresis loss per cycle per unit volume   =...ergs/cycle/cm^3

Precautions and Sources of Errors:

1. Intensity of light on the screen of C.R.O. should not be large to avoid the burning of screen.
2. The current flowing in the primary coil of the solenoid should be sufficiently large so that the magnetizing field produced should be enough to magnetise the specimen fully.
3. The C.R.O. should be carefully handled.

Viva-Voce

Q.1. What are you doing?

Ans. Sir/Madam, I am doing drawing hysteresis curve of a specimen in the form of a tranformer on C.R.O.

Q.2. What do you mean by hysteresis curve?

Ans. A curve showing the variation of magnetic flux density B as a function of magnetising field H is called magnetisation curve. A plot of B versus H of a ferromagnetic material in which the material is magnetised in one direction and then in opposite direction is called the hysteresis curve of the specimen.

Q.3. What do you mean by hysteresis loss?

Ans. When a magnetic material is magnetised by applying some external magnetic field the energy is absorbed by the sample. When the external field is removed or when the magnetic material is demagnetised the energy absorbed by the material during magnetisation is not completely recovered. Thus, the sample retains some energy. The energy remaining in the sample is lost as heat. This loss of energy in the form of heat is called hysteresis loss.

Q.4. What is the intensity of magnetisation?

Ans. The intensity of magnetisation I of a magnetic field is defined as the magnetic moment per unit volume.

Q.5. What do you mean by magnetic intensity ( )

Ans. The capability of the magnetic field to magnetise a material place in it, is expressed by means of a magnetic vector H, called the magnetic intensity of the field.

Q.6. What is the cause of hysteresis ?

Ans. In the processes of magnetisation of a magnetic materia some energy is absorbed by the substance, but this absorbed energy is not totally recoverable due to reten.tivity. Hence, retentivity of the material causes hysteresis.

Q.7. What type of iron cores are suitable for electromagnets, transformers, chokes and telephone diaphragms ?

Ans. For cores of transformer, armature of dynamos, motors, A.F. chokes and telephone diaphragms, the material should have high initial permeability for low value of H and a high specific resistance for reducing eddy current losses.

Apart from it, low hysteresis loss is also an important factor because core materials are continuously subjected to a cyclic change. Soft iron is suitable for this purpose.

Q.8. What is hysteresis ?

Ans. The phenomenon in which the magnetic induction B lags behind the magnetising field H during a magnetic cycle is called hysteresis.

Q.9. Why permanent magnets are made of steel ?

Ans. Because the coercivity of steel is much greater than that of iron so permanent magnets are made of steel.

Q.10. What do you mean by coercivity ?

Ans. The coercivity of a material is a measure of the strength of the reverse magnetising field required to wipe out the residual magnetism.

Experiment No. 14

Object: To determine ballistic constant K, of a moving coil ballistic galvanometer with the help of a standard condenser of known capacity.

Apparatus required: Ballistic galvanometer, damping key, rheostat, morse key, accumulator, a condonser of known capacity, voltmeter and connection wire.

Description of Apparatus and Circuit: The circuit arrangement for the determination of ballistic constant K with the help of a standard condenser of known capacity (of about 0.05μF) is shown in the given Fig.14.1. In this arrangement, a battery E in series with a key K_1 is connected to the lower terminals P and Q of a

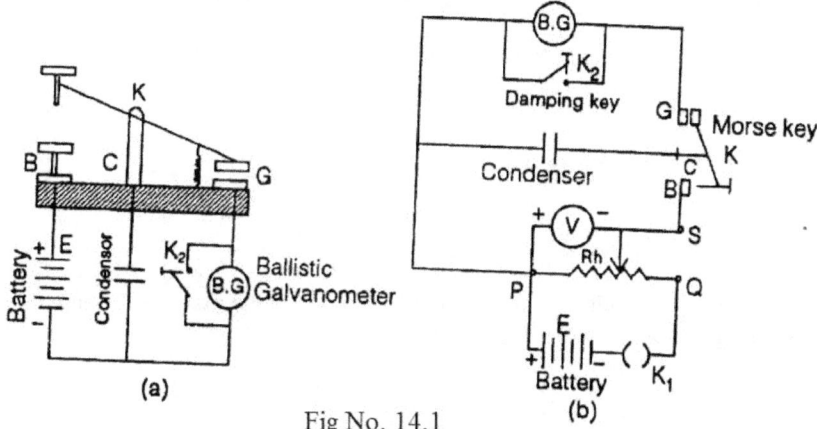

(a)   (b)

Fig No. 14.1

rheostat Rh. To use rheostat as a potential divider, the sliding contact terminal S is connected to the terminal B of the morse key and the lower positive terminal P to the one end of the ballistic galvanometer B.G. The terminal P is also connected to the one end of the standard condenser C whose other end is connected to the common terminal C of the morse key. The remaining terminal G of the morse key is connected to the remaining teminal of the ballistic galvanometer. To stop the motion of the coil of the ballistic galvanometer a tapping key (or damping key) K_2 is connected across its ends. Between the sliding contact terminal S and fixed terminal P of the rheostat a voltmeter V is connected to read the voltage used for charging the condenser C.

Theory and Formula Used: When a transient current or charge passes through a ballistic galvanometer it gives an angular impulse to the coil which in turn begins to oscillate about its own axis.

The ballistic constant k of the galvanometer is given by

$$- = \frac{}{-\frac{\cdot}{2}(1 + -)}$$

where

C = capacity of the condenser in farads,

E = voltage of the cell used for charging the condenser (in volts),

$\mathfrak{z}\mathfrak{t}$ ι= first observed throw the condenser is discharging through the galvanometer, and

$\mathfrak{z}\mathbf{i}$ = logairthmic decrement

The logairthmic decrement is given by

$$\frac{}{\mathfrak{l}\prime} = \frac{2.3026}{10} \times \log \quad \underline{}$$

If a lamp and scale arrangment is used and d_1 and d_{11} are the first and eleventh throws of spot of light on the scale placed at distance D then

$$\frac{\cdot 1}{\mathfrak{n}} = \div \quad \text{and} \quad \frac{\text{м}}{\mathfrak{n}} = -\!-$$

Therefore,

$$\frac{\theta}{\theta_1} = \frac{d}{d}$$

Then $\dfrac{\prime}{\mathfrak{z}} = \dfrac{1}{\mathfrak{c}} \times \log \quad - \cdot$

Procedure:

1. Press the morse key in such a way that the condenser is connected to the cell and thereby charging of the condenser takes place.
2. Release the morse key so that the condenser is connected to the galvanometer and thereby discharges.
3. Note the first and eleventh throw distance of the galvanometer spot on the scale and the corresponding reading of the voltmeter.
4. Repeat the procedure for different values of applied voltage.

Observations:

C =........Farad

S.No.	Capacity C of condensor	Applied voltage E volt	Final position of the light spot corresponding To throw of galvanometer		$=\dfrac{2.3026}{10}$ $\times \log \dfrac{e a l}{\lambda}$	Mean
			First throw ( )mm	Eleventh throw ( )mm		
1.						
2.						
3.						
4.						

Calculation: The logarithmic decreament λ is calculate by

$$\frac{\lambda}{\lambda} = \frac{1}{\lambda} \times \log -\ -\ \cdot\ = \ldots\ldots\ldots\ldots$$

The ballistic constant K is calculated by

$$K = \frac{}{(-)} = \ldots\ldots\text{coulomb/mm}$$

(Calculated separately for each set of observation then find its mean value).

Result: The ballistic constant (K) of the given galvanometer = ...coulomb/mm

Precautions and Sources of Error:

1. The ballistic galvanometer should be properly levelled so that the coil becomes free to rotate symmetrically in the space between the pole pieces of the electromagnet without touching the pole pieces and core of the coil.
2. A damping key should always be connected across the terminals of the galvanometer for overcoming unwanted oscillatory motion of the coil.
3. A storage battery of steady e.m.f. is used.
4. Galvanometer coil should be made stationary for each set of observations before the condenser is discharged through it.
5. All the connecting terminals should be well tightened.

Viva- Voca

Q.1. What are you doing?

Ans. Sir/Madam, I am determining the ballistic constant of a ballistic galvanometer with the help of a standard condenser.

Q.2. What do you mean by ballistic constant of a galvanometer?

Ans. It is a constant (K), which when multiplied by the first throw of the galvanometer gives the amount of charge passing through its coil. Mathematically it is defined as

$$q = K\theta$$

Q.3. On what factors does the ballistic constant depends?

Ans. The ballistic constant K is directly proportional to the time period T of oscillation of the coil and the restoring couple per unit twist C of the suspension fibre. It is inversly proportional to the strength (B) of the electromagnet in which its coil rotates, number of turns on the coil (n) and to the area of cross-section A of the coil. Mathematically ballistic constant is written as

$$K = \frac{.\quad}{} \ .$$

Q.4. What is the unit of ballistic constant?

Ans. If the deflection is measured in mm. then the unit of ballistic constant K is coulomb/mm. On the other hand its unit is coulomb/radian if the throw is measured in radian.

Q.5. What are the essential feature of a ballistic galvanometer?

Ans. The essential features of a ballaistic galvanometer are as follows:

1. A copper wire coil which is wound on a non-conducting frame.
2. The moment of inertia of the coil should be large and its suspension has a small torsional constant.

Q.6. What are the characteristics of a ballistic galvanometer?

Ans. The general characteristics of a ballistic galvanometer are as follows:

1. Its time period is large
2. Damping is very small
3. Whole of the charge passes through the coil before it moves from its position of rest.

Q.7. What is the practical unit of a capacity? Define it.

Ans. The practical unit of a capacity is Farad. The capacity of a condenser is 1 farad if the charge of one coulomb produces a potential difference of 1 Volt in it.

Q.8. What is the principle of a condenser.

Ans. When an insulated conductor is brought near a charged conductor, the capacity of a charged conductor increases. The increase in capacity is much more if the insulated conductor be earthed.

Q.9. What is a capacitor or a condenser ?

Ans. A capacitor or condenser is an arrangement which can store sufficient quantity of electric charge. It consists of two neighbouring conductors that have equal and opposite charges.

Q.10. Why do you get deflection ?

Ans. On changing the direction of current in the primary, the magnetic flux changes in the secondary and due to the induced charge passing through the galvanometer the deflection is produced in its coil.

Experiment No. 15

Object: To determine coefficient of viscosity of water by Meyer's disc method.

Apparatus required: Meyer's apparatus(Torsional pendulum), lamp and scale arrangement, stop clock, given liquid in a container, etc.

Description of Apparatus: Meyer's apparartus consists of a flat circular disc suspended from its center by uniform wire. The top end of the wire is fixed firmly in a vertical chuck. The length of suspension wire can be adjusted by working on the chuck. Just above the center of disc, a cylindrical rod is supported on the wire symmetrically. A small circular concave mirror is attached at the mid point of the rod.

A lamp and scale arrangement is placed in front of concave mirror at suitable distance from it (approximately 1 meter). The scale is adjusted so that light from the lamp after reflection by concave mirror forms a circular spot with cross wire at the mid zero division of the scale. The scale is calibrated in centimetres on either side of the zero.

Formula used:

The coefficient of viscosity of water by logarithmic decrement method is given by:

$$\eta = \frac{1}{+2\pi \cdot d^2 \cdot \left(\frac{1}{T_0} \right)} \left[1 + \frac{i-i}{-} + \text{ }^{12} \right]$$

where

I = Moment of inertia of torsional pendulum about the axis of suspension,

T_0 = Period of oscillation of the system in air,

ρ = The density of wtaerat room temperature,

r = Radius of the disc,

d = Thickness of the disc,

λ_0 = Logarithmic decrement in air, and

λ = Logarithmic decrement in water.

Thus coefficient of viscosity of water can be determined by finding the moment of inertia of torsional pendulum in first part and in the second part logarithmic decrements in air and water must be determined.

1. Moment of inertia (I) by torsional oscillations with symmetrical masses:

The moment of inertia of torsional pendulum given by,

$$I = -2\pi r \times 2 \cdot \left(\frac{4}{1}\right) - \quad)$$

where T_0 is the period of oscillation without masses. T_1 is the period of oscillations when two symmetrical identical bodies are kept at a distance d_1 from the center of the disc on either side of it, T_2 is the period of oscillations when two bodies are kept at distance d_2 from the center of the disc and m is the mass of each symmetric body.

2. Logarithmic decrements in air (λ_0) and in water (λ):

The logarithmic decrement in air is given by the formula,

$$\lambda_0 = \frac{1}{\eta_1} \log \frac{\cdots\cdots\cdots\alpha\cdots\cdots}{\alpha_{n-2}\cdots\cdots\alpha_{1}+\cdots} \quad -$$

where, $\eta_1 = 2T_a/T_0$; T_a is the time taken by the disc to reduce its full deflection half of its value. a_1, etc. are mean (mean of left and right) full deflections and α_{n+1}, etc. are mean (mean of left and right) half deflection due to damping of the disc in air.

The logarithmic decrement in water is given by

$$\lambda = \frac{1}{\eta_2} \log \frac{\cdots\cdots\cdots\beta\cdots\cdots}{\beta_{n-2}\cdots\cdots\beta_{1}+\cdots} \quad -$$

where $\eta_2 = 2T_w/T_0$; T_w is the time taken to reduce its full deflection into half deflection of the disc in water. Here β, etc. are mean full deflection, β_{n+1} etc. are mean half deflections due to damping of the disc in water.

Procedure:

1. To determine moment of inertia(I):

The lamp and scale are adjusted to get the light spot at zero division of the scale, when torsion pendulum is in equilibrium. A small twist is given to the disc so as to set it in torsion oscillation. The time for ten oscillations is noted with two trials by observing the light spot oscillating to and fro. The period of oscillation T_0 is determined. Now, two symmetrical bodies, each of mass m are placed closest to the center of disc on either side of it. The distance between two masses $2d_1$ is measured. The pendulum is set into torsional oscillations, the time for ten oscillation and hence the period of oscillations T_1 is determined. Next, masses are kept farthest from the center at a distance $2d_2$ from each other. Experiment is repeated and period of oscillation T_2 is determined. Readings are recorded as in Table A.

2. To determine logarithmic decrements in air and water:

The disc is set in torsional oscillations. Leaving first few oscillations, when the spot is at extreme end, say, left end on the scale, stop clock is started and the scale reading is noted. Observation is repeated in taking successive ten readings on either extremen end of the scale. The average of first left and right readings gives α_1 and so the other readings, as tabulated in Table B. The total time taken to T_a to reduce the first extreme left reading to half of its value is noted and once again ten half deflections on either side are noted and the mean values α_{n+1} etc are found out. Using the formula, logarithmic decrement λ_0 is calculated.

Now to determine logarithmic decrement λ in water, the disc is immersed in a basin of water (not to touch wall or bottom of basin) and lamp and scale are adjusted to see the light spot at zero division when the pendulum is in equilibrium. Since the viscosity of water is large as compared with air, a slightly different method is adopted to determine λ. Starting from first extreme left reading, stop clock is started and readings are noted on both sides successively. The time required T_w to reduce the first full deflection into half is noted and readings are recorded as given in Table C. Using the formula, logarithmic decrement λ is determined.

The radius r of the disc is calculated by measuring its circumference, $2\pi r$.

Observations:

Table A - Period of Oscillations

Disc	Time for 10 oscillations		Mean (S)	Period of oscillation
	I	II		
Without masses				T_0
Masses at d_1				T_1
Masses at d_2				T_2

Table B - Logarthmic Decrement λ_0: T_a = ...sec.

Scale readings full deflection (cm)		Mean	Scale readings half deflection (cm)		Mean
Left	Right		Left	Right	
α_{IR}	α_{IR}	α_I	$\alpha_{IL}/2$	$\alpha_{IR}/2$	α_{n+1}

Table C - Logarithmic Decrement λ : T_w = ...sec

Scale readings full deflection (cm)		Mean	Scale readings half deflection (cm)		Mean
Left	Right		Left	Right	
β_{IL}	β_{IR}	β_I	$\beta_{IL}/2$	$\beta_{IR}/2$	β_{n+1}

Calculations:

The viscosity of water is calculated by following formula

$$\eta = \frac{1}{+2\tau \cdot (l)^2 \cdot \left(\eta \right)} \left[+ \frac{\lambda - \lambda}{\cdot} + \frac{\lambda 2}{\cdot} \right.$$

= ....poise (at 20°C)

Result:

The viscosity of water = ...poise (at 20°C)

Standard result: The standard value of viscosity of water at 20°C = 0.01004 poise

Percentage Error: The percentage error in the experimental value is calculated by the following formula

$$\text{Percentage error} = \frac{\text{d value} - \text{Calculated value}}{\text{d value}} \times 100 \times 100\%$$

$$= ....\%$$

Precautions and Sources of Error:

1. The disc should be immerse in basin of water such that it does not touch the wall and the bottom of basin.
2. The disc should be set I torsional oscillations.
3. The container should contain suitable quantity of water for this experiment.

VIVA VOCA

Q.1. What are you doing?

Ans. Sir/Madam, I am determining the viscosity of water by Meyer's disc method.

Q.2. What is viscosity ?

Ans. In the presence of a relative motion between two layers of a liquid, an opposing tangential force sets in between the layers to destroy the relative motion. This property of the liquid is termed viscosity and is analogus to friction.

Q.3. What is the coefficient of viscosity?

Ans. The tangential viscous force acting per unit area over two adjacent layers of the liquid for a unit velocity gradient is referred to as the coefficient of viscosity.

Q.4. How does the coefficient of viscosity changes with temperature?

Ans. The coefficient of viscosity decreases with rise in temperature in case of liquids, but for gases it increases with rise in temperature.

Q.5. Can you use this method for all types of liquids?

Ans. No this method can be suitably applied for liquids of low viscosity. For highly viscous liquids, Stokes's method can be used.

Appendices

Physical Constants and Their Standard Values in Tabular Form

Table I	: Some Fundamental Constants
Table II	: Wavelength of Some spectral Lines (Å)
Table III	: Refractive Index of Some solids and Liquids
Table IV	: Refractive Index and Dispersive power of some Solids, Gases,Liquids and crystals
Table V	: Specific Rotation of Some Solutions and pure Liquids
Table VI	: Specific Resistance and Temperature Coefficient of Resistance of some Metals and Alloys
Table VII	: Internal Resistance and E.M.E. of Some Primary and Secondary Cells
Table VIII	: Wire Resistance of Various Standard Wire Gauges (S.W.G.)
Table IX	: Hall constants-Energy Band Gap E_g, Conductivity σ and Mobility μ_n and μ_p for some Materials
Table X	:Hall constants
Table XI	: Magnetic Elements at Various Places
Table XII	: Electro-Chemical Equivalent of Various Metals
Table XIII	: Viscosity and Surface tension of Some Common Liquids
Table XIV	: Viscosity and Surface Tension of Water
Table XV	: Ionisation Potential of Some Gases
Table XVI	: Acceleration Due to Gravity at Various Places on Earth
Table XVII	: Elastic Constants of Some Metals and Alloys
Table XVIII	: Densities of some Solids, Liquids and Gases
Table XIX	: Thermal Constants of Important Solids, Alloys and Liquids
Table XX	: Electromagnetic Spectrum (Wavelengths)
Table XXI	: Transistors Manufactured by Bharat Electronics Limted (BEL)
Table XXII	: Diodes Manufactured by BEL

• **Colour Code for High Resistances Used in Labs**

• **Symbols Applicable in the Circuits of Electricity and Electronics**

Physical Constants and their Standard Value in Tabular Form

Table I: Some Fundamental Constants

Constant	Symbol	Computational value
Speed of light in a vaccum	c	3.00×10 m/s
Elementary charge	e	1.60×10 C
Electron mass	m_e	9.11×10 kg
Proton mass	m_p	1.67×10 kg
Ratio of proton mass to electron mass	m_p/m_e	1840
Neutron mass	m_n	1.68×10 kg
Muon mass	m_μ	1.88×10 kg
Electron mass	m_e	5.49×10 u
Proton mass	m_p	1.0073 u
Neutron mass	m_n	1.0087 u
Hydrogen atom mass	m_{1H}	1.0078 u
Deuterium atom mass	m_{2H}	2.0141 u
Helium atom mass	m_{4He}	4.0026 u
Electron charge-to-mass ratio	e/m_e	1.76×10 C/kg
Permittivity constant	ε_0	8.85×10 F/m
Permeability constant	μ_0	1.26×10 H/m
Planck constant	h	6.63×10 J.s
Electron Compton wavelength	λ_C	2.43×10 m
Universal gas constant	R	8.31 J/mol.K
Avogadro constant	N_A	6.02×10 mol^{-1}
Boltzmann constant	k	1.38×10 J/K
Molar volume of ideal gas at STP	V_m	2.24×10^{-1} m^3/mol
Faraday constant	F	9.65×10 C/mol
Stefan-Boltzmann constant	σ	5.67×10^2 W/m^2. K^4
Rydberg constant	R	1.10×10 m^{-1}
Gravitational constant	G	6.67×10 m^3/s^2.kg
Bohr radius	r_B	5.29×10 m
Electron magnetic moment	μ_e	9.28×10 J/T
Proton magnetic moment	μ_p	1.41×10 J/T
Bohr magneton	μ_B	9.27×10 J/T
Neuclear magneton	μ_N	5.05×10 J/T
Acceleration due to gravity	g	9.8 m/sec^2
Mechanical equivalent of heat	J	4.18 J/sec

Table-ll

Wavelength of Some Spectral Lines (A°)

Wavelength of prominent spectral lines (A°) of some common gases or vapours. Primary standard : Red Cadmium line 6438.4696 A°

$$(1 \text{ A}° = 10^{-8} \text{ cm})$$

Hydrogen		Cadmium	Mercury
(c)6562.784 (α) r		6438	6908 R
(F) 4861.327 (β) gb		5086	6234 O
4340.466 (r) b		4800	5791 Y
4101.736 (δ)v		4678	5770Y
		4662	5461g
Sodium			4960g
5890	D₁ 0	Helium	4916 b, g
5896	D₂ 0	7065 R	4358 b, g
		6678 R	4348 b
Lithium		(D₃) 5876 Y	4339 b
6104	O	4471 b	4078 y
6708	R	4026 v	4047 v
		3889 v	Neon
Potassium			7245 R
4047	v		6507 R
4703	b		6402 R
5452	g		6334 R
5912	y		6266 R
6031	O		5945 O
7700	r		5882 Y
			5852 Y
			5342 g
			5341 g
			5331 b, g.

Table-lll
Refractive Index of Some Solids and Liquids

Refractive Index of Substances (at 15^0 C for D-line of Sodium relative to air for $\lambda = 5893A^\circ$)

Substances	R.I. (μ)	Substance	R.I. (μ)
Isotropic Solids		Liquids	
Calcite	1.658 O-ray		
	1.486 E-ray		
Diamond	2.417	Aniline	1.590
Ice	1.31	Cedar oil	1.516
Mica	1.56 – 1.69	Benzene	1.504
Rock salt (NaCl)	1.544		
Sugar	1.56	Canada Balsam	1.53
Silvine (KCl)	1.490	Chloroform	1.53
Topaz	1.63	Ethyl Alcohol	1.33-1.41
Quartz	1.544 O-ray	Glycerine	1.449
	1.553 E- ray	Kerosene oil	1.39
Glass :		Sulphuric	1.47
Crown	1.500	Ether	1.354
Dense Crown	1.620	Turpentine	1.43
Flint	1.560	Olive oil	1.46
Soda	1.50	Water	1.333
Dense Flint	1.620	Paraffine oil	1.44
Extra dense flint	1.650		
Very dense flint	1.720		

Table-IV
Refractive Index and Dispersive Power of some Solids, Gases, Liquids and Crystals
μ= Refractive index for $\lambda = 5893A^0$ (D-line)
ω = Dispersive power referred to F (4871 A^0), D(5193A^0) and C (6563A^0) lines.

Substance	μ	Ω	Substance	μ	ω
Isotropic solid			Liquids		
Canada Balsam	1.530	-	Alcohol (Ethyl)	1.362	0.017
Diamond	2.425	-	Alcohol (Methyl)	1.329	0.016
Fused Quartz	1.4584	0.015	Benzene	1.501	0.033
Fused Silica	1.4585	0.0148	CCl_4	1.46	0.020
Ice	0.31	-	Chloroform	1.446	0.010
NaF	1.325	0.0118	Ether	1.354	0.07
LiF	1.392	0.0101	Glycerol	1.478	-
CaF (Fluorite)	1.434	0.0105	Quinoline	1.627	0.050
KCl (Sylvine)	1.490	0.0027	Turpentine Oil	1.47	-
NaCl (Rock Salt)	1.5443	0.0234	Water	1.333	0.018
KI	1.667	0.0431			
			Uniaxial Crystals		
			Calcite	1.581(O)	
Gases				1.575(E)	
Air (N.T.P.)	1.000292	-			
			Quartz	1.544(O)	0.0143
				1.552(E)	0.0144
			Calspar	1.658(O)	0.0203
				1.486(E)	0.0125

Table –V
Specific Rotation of Some Solutions and Pure Liquids

Optically active substance	Solvent	Specific Rotation
Camphor	Alcohol	$+41.0^0$
Cane Sugar	Water	$+66.5^0$
Fructose	Water	-91.0^0
Glucose	Water	$+52.5^0$
Invert Sugar	Water	-19.5^0
Nicotine	Pure	-122.0^0
Tartaric scid	Water	$+8.9^0$
Turpentine	Pure	-192.0^0

Table-VI

Specific Resistance and Temperature Coefficient of Resistance of Some Metals and Alloys

Substance	Composition	Specific resistance ohm x cm x 10^{-6}	Temperature coefficient of resistance per ^0C x 10^{-4}
Aluminium	--	2.8	39
Brass	70% Cu, 30%Zn	6.6	10
Copper	--	1.78	42.8
Constantan	60% Cu – 40%Zn	49.1	-0.4 to + 0.1
Manganin	84% Cu, 4% Ni, 12% Mn	44.5	0.02 – 0.5
Mercury	--	95.8	8.9
Nichrome	80% Ni, 20% Cr	110.0	1.7 – 3.5
Platinum	--	11.0	37.0
Silver	--	1.65	40.0
German silver	62% Cu, 75% Ni, 22% Zn	28	2.3 – 6.0

Table-VII

Internal Resistanace and E.M.F. of Some Primary andSecondary Cells

Name	Internal Resistance (ohms)	E.M.F.(volts)
Cadmium cell	900Ω (very high)	1.0183 V at 20^{0}C
Daniel cell	3 Ω- 4 Ω (fairly constant)	1.080
Leclanche cell	Fairly high and increases with use	1.46
Standard clark cell	500-1000 Ω (very high)	1.4328 at 15^{0}C
Alkali Accumulator	Low	1.35
Lead Accumulator	Very low	2.10

Table – VIII

Wire Resistance of Various Standard Wire Gauges (S.W.G.)

S.W.G.No	Diameter (mm)	Resistance (ohm/meter)			
		Copper	Constantan (60% Cu, 40% Ni)	Manganin (84% Cu, 4% Ni, 12% Mn)	German Silver
10	3.25	0.0021	0.057	0.051	0.049
12	2.64	0.0032	0.086	0.077	0.041
14	2.03	0.0054	0.146	0.131	0.070
16	1.63	0.0083	0.228	0.205	0.109
18	1.22	0.0148	0.495	0.361	0.193
20	0.914	0.0260	0.722	0.645	0.345
22	0.711	0.0235	1.200	1.07	0.57
24	0.559	0.070	1.930	1.73	0.92
26	0.457	0.105	0.890	2.58	1.38
28	0.374	0.155	4.270	3.82	2.02
30	0.315	0.222	6.080	5.45	2.90
32	0.274	0.293	8.020	7.18	3.83
34	0.234	0.404	11.100	9.9	5.27
36	0.193	0.590	16.200	14.5	7.74
38	0.152	0.950	26.200	23.2	1.24
40	0.122	1.480	40.600	36.3	1.94
42	0.102	2.100	58.500	53.4	-
44	0.081	3.300	91.400	81.7	-
46	0.061	5.900	162.500	145.5	-

Table – IX

Hall Constants-Energy Band Gap E_g, Conductivity σ and Mobility μ_n and μ_p for Some materials

Substance	Energy band gap E_g (eV)	Conductivity σ (mho – cm^{-1})	Mobility μ (cm^2/volts-sec)	
			μ_n	μ_p
Silicon	1.12	23 X 10^4	1350	480
Germanium	0.67	47	3900	1900
Diamond	7.2	10^{12}	1800	1200
Tellurium	0.38	21 x 10^{-6}	300	1170

Table-X
Hall Constants

Material	Hall Coefficient (R_H)	No. of Charge Carriers (n)	Hall angle (φ)	Mobility (μ)
Graphite mixed carbon (used in Torchcells)	1.905 ×10^2 e.m.u	3.28×10^{17}	0.036 rad	502 cm^2/volt-sec

Table-XI
<u>Magnetic Elements at Various Places</u>

Station	Declination	Angle of Dip	Horizontal component H (oersted)	Vertical Component V (oersted)
Agra	0^0 10' E	40^0 40'	0.3484	0.2985
Ajmer	0^0 10' E	39^0 20'	0.3483	0.2861
Aligarh	0^0 20' E	$41^0$50'	0.3455	0.3091
Allahabad	0^0 20' W	$37^0$10'	0.3629	0.2758
Bareilly	0^0 20' W	42^0 20'	0.3436	0.3136
Mumbai	0^0 00'W	25^0 30'	0.3761	0.1792
Kolkata	0^0 00' W	31^0 30'	0.3819	0.2342
Chandausi	0^0 30' E	42^0 40'	0.3429	0.3167
Dehradun	0^0 00' E	45^0 50'	0.3315	0.3405
New Delhi	0^0 40' E	42^0 52'	0.3453	0.3204
Gorakhpur	0^0 20' W	30^0 40'	0.3576	0.2972
Gwalior	0^0 20' E	39^0 00'	0.3531	0.2857
Jaipur	0^0 30' E	40^0 30'	0.3470	0.2961
Jodpur	0^0 00'	39^0 10'	0.3482	0.2835
Kanpur	0^0 00'	38^0 39'	0.3628	0.2901
Khurja	0^0 30' E	42^0 10'	0.3426	0.3109
Lucknow	0^0 10' W	40^0 00'	0.3536	0.2965
Meerut	0^0 40' E	43^0 30'	0.3389	0.3212
Udaipur	0^0 00'	35^0 50'	0.3620	0.2613
Varansi	0^0 30' W	37^0 10'	0.3635	0.2764

Table-XII
Electro Chemical Equivalent of Various Metals

Element	Atomic Weight	Valency	E.C.E.(gm./coulomb)
Aluminium	26.97	3	0.0000935
Copper	63.57	2	0.0003294
Gold	197.20	3	0.0006809
Lead	103.61	2	1.0010731
Nickel	58.69	2	0.0000304
Oxygen	16.00	2	0.0000829
Silver	107.88	1	0.0011180
Zinc	65.38	2	0.0003383

Table-XIII
Viscosity and Surface Tension of Some Common Liquids
Converson Factor : [10 Poise = 1 kg m^{-1} s^{-1}], [1gm/cm^3 = 10^3 kg/m^3]

Name	Density at 20^{0}C (gm/cm^3)	Viscosity at 20^{0}C (Poise)	Surface tension at 20^{0}C	
			(dynes/cm)	(10^{-2} Newton/m)
Benzene	0.9	648 ×10^{-5}	28.9	2.89
Chloroform	1.525	563 × 10^{-5}	27.15	2.71
Ethyl Alcohol	0.79	1190 ×10^{-5}	22.31	2.23
Glycerin	1.26	83	62.5	6.25
Methyl Alcohol	0.807	590 × 10^{-5}	22.5	2.25
Turpentine oil	0.873	1490 ×10^{-5}	27.2	2.72

Table-XIV

Viscosity and Surface Tension of Water

Temperature	0^0C	10^0C	20^0C	30^0C	40^0C	50^0C	60^0C	70^0C	80^0c	90^0C	100^0C
Viscosity (poise)	.01793	.01308	.01004	.00799	.00657	.00550	.00469	.00406	.00356	.00316	.00289
Surface Tension (dynes/cm)	75.6	74.21	72.75	71.19	69.58	67.3	66.1	63.8	62.0	60.2	58.2

Table-XV
Ionisation Potential of Some Gases

Name of the gas	Ionisation Potential (Volts)	Name of the gas	Ionisation Potential (Volts)
Argon	15.75	Neon	21.56
Helium	24.56	Oxygen	13.60
Hydrogen	13.58	Sodium	5.14
Mercury	10.40	Xenon	12.00

Table – XVI
Acceleration Due to Gravity at Various Places on Earth

Place	g (cm/sec^2)	Place	g (cm/sec^2)
Agra	979.06	Khurja	979.66
Ajmer	978.60	Lucknow	979.06
Aligarh	978.08	Chennai	978.25
Allahabad	978.95	Mathura	979.09
Bareilly	979.17	Meerut	979.15
Mumbai	978.65	Moradabad	979.19
Kolkata	978.78	Nainital	978.70
Dehradun	979.07	Rampur	979.19
New Delhi	979.15	Saharanpur	979.19
Gorakhpur	979.05	Udaipur	979.68
Gwalior	978.97	Varanasi	978.99
Indore	978.60	Equator	978.03
Jaipur	978.52	Pole	982.22
Kanpur	979.01		

Table – XVII
Elastic Constants of Some Metals and Alloys

(Approximate Values)

Substance	Young's Modulus $Y \times 10^{11}$ (dyne/cm^2)	Rigidity $\eta \times 10^{11}$ (dyne/cm^2)	Bulk Modulus $K \times 10^{11}$ (dyne/cm^2)	Poisson's Ratio σ	Tensile Strength (wires)$\times 10^9$ (dyne/cm^2)
Metals					
Aluminium	7.05	2.62	7.58	0.345	2.0-4.5
Copper	12.98	4.83	13.76	0.343	2.8-4.6
Gold	7.8	2.7	21.07	0.44	-
Iron (wrought)	19.20	7.7-8.3	16.9	0.29	-
Iron (cast)	10.13	3.5-5.3	9.6	0.23-0.31	6.9
Lead	1.62	0.56	4.6	0.441	0.12-0.17
Nickel	20.4	7.9	16.1	0.28	5.0-9.0
Platinum	16.8	6.1	22.8	0.377	3.3-3.7
Silver	8.27	3.03	10.4	0.367	2.9
Steel(cast)	21.0	8.10	16.88	0.293	-
Steel (mild)	20.21	7.9-8.9	16.8	0.25-0.33	11-23
Alloys					
Brass	9.7-10.2	3.3-3.7	11.2	0.34-0.40	3.5-5.5
Constantan	16.3	6.11	15.7	0.327	-
German Silver	11.6	4.3-4.7	-	0.37	4.6
Invar	17.6	-	-	0.11-0.27	-
Manganin	12.4	4.65	12.4	0.334	
Phosphor-bronze	12.0	4.36	-	0.38	10.8
Glass (crown)	5.0-7.8	2.6-3.2	-	0.20-0.27	-
Glass (Flint)	5.0-6.0	2.3	4.0-5.0	0.2-0.26	-
Quartz Fibre	7.3	3.1	3.7	0.77	0.3-0.9
Fused Rubber	0.0015 - 0.0005	0.00005 - 0.00015	3.7 -	0.46-0.499 -	-10 -

Table- XVIII
Densities of some Solids, Liquids and Gases

Substances	Density (gm/cm³)	Substances	Density (gm/cm³)
SOLIDS:		Platinum	21.45
Aluminium	2.7	Platinum	21.45
Antimony	6.62	Quartz	2.66
Asbestos	2.0-2.8	Silver	10.5
Bismuth	9.78	Selenium	4.8
Brass	8.4-8.7	Sodium	0.97
Bronze	8.8-8.9	Tin	7.3
Carbon	2.22	Tungsten	19.3
Chromium	6.92	Zinc	7.1
Constantan	8.88	Wax (parafin)	0.875
Copper	8.89	**GASES:**	
Cork	0.22-0.26	Air	0.00129
Germanium	5.3	Carbon dioxide	0.00198
Glass crown	2.0-2.6	Helium	0.000179
Flint	2.9-4.4	Hydrogen	0.00609
Pyrex	2.25	Steam	0.00091
Gold	19.3	**LIQUIDS:**	
Ice	0.9167	Alcohol	0.80
Iron, pure	7.88	Aniline	1.02
Wrought	7.84	Benzene	0.88
Cast	7.6	Ether	0.736
Steel	7.7	Glycerine	1.26
Lead	11.34	Lubricating oil	0.91
Magnesium	1.74	Mercury	13.60
Mica	2.6-3.2	Methyl alcohol	0.83
Manganin	8.50	Turpentine	0.87
Nickle	8.85	Water (0^0C)	0.9982

Table –XIX
Thermal Constants of Important Solids, Alloys and Liquids

Substance	Melting p.t (^{0}C)	Boilling p.t (^{0}C)	Specific heat	Latent heat	Coefficient of linear expansion	Thermal conductivity (Cal/sec/cm/^{0}C)
Aluminium	658	1800	0.22	92.4	25×10^{-6}	0.504
Bismuth	269	1560	0.03	13.4	-	0.0194
Brass	905	-	0.09	-	19	0.26
Copper	1084	2360	0.93	43	16.3	0.918
Constantan	-	-	0.098	-	-	0.054
Ebonite	-	-	0.40	-	-	0.00038
Gold	1063	2360	0.032	16	-	0.7
Glass	-	-	0.16	-	-	0.0025
Ice	0	100	0.5	79	50.7	0.0052
Iron (wrought)	1530	2450	0.12	49	-	0.144
Lead	327	1755	0.031	5	29.1	0.083
Nickel	1453	-	0.109	-	13	0.142
Platinum	1774	4300	0.032	27	8.8	0.166
Silver	961	2152	0.056	22	19.2	0.974
Tungsten	3387	4830	0.03	-	-	0.35
Steel	1440	-	0.11	-	11.0	0.115
Zinc	420	-	0.092	-	26.3	0.265
Cork	-	-	0.49	-	-	0.00012
Rubber	-	-	0.38	-	70.4	0.00045
Benzene	5.5	80.2	0.34	95	-	3.3×10^{-4}
Ether	-132	34.6	0.56	88	-	3.1×10^{-4}
Glycerine	17	-	0.58	-	-	6.37×10^{-4}
Mercury	-38.9	357	0.033	68	-	-
Turpentine	-10	-	0.42	-	-	3.25×10^{-4}
Water	0	100	1.00	539	-	1.47×10^{-4}

Table- XX
Electromagnetic Spectrum (Wavelengths)

Wireless Waves	5 meters and above
Infra red	3×10^{-2} cm to 7.5×10^{-5} cm
Visible red	7.5×10^{-5} cm to 6.5×10^{-5} cm
Visible Orange	6.5×10^{-5} cm to 5.9×10^{-5} cm
Visible Yellow	5.9×10^{-5} cm to 5.3×10^{-5} cm
Visible Green	5.3×10^{-5} cm to 4.9×10^{-5} cm
Visible Blue	4.9×10^{-5} cm to 4.2×10^{-5} cm
Visible Violet	4.2×10^{-5} cm to 3.9×10^{-5} cm
Ultra Violet	3.9×10^{-5} cm to 1.8×10^{-5} cm
Soft X-rays	2.0×10^{-7} cm to 1.0×10^{-8} cm
Hard X-rays	1.0×10^{-8} cm to 1.0×10^{-9} cm
Υ-rays	5.0×10^{-9} cm to 5.0×10^{-10} cm
Cosmic rays	5.0×10^{-12} cm

Table- XXI
Transistor Manufactured by Bharat Electronics Limited (BEL)

No.	Type
PNP	AC125
PNP	AC126
PNP	AC127
PNP	AC128
PNP	AC134
PNP	AF115
PNP	AF116
PNP	AF117

Table- XXII
Diodes Manufactured by BEL

OA 70	OA 73	OA 72	OA 79	OA 81	OA 85	OA 91	OA 95	DR 25	DR 100

<u>Colour Code for High Resistance Used in Labs</u>

Generally, carbon resistance are provided in the laboratory for radio works. The value of these resistance is given in terms of colour code in which the numbers from 0 to 9 are represented by the following colours:

Numbers	Colour
0	Black
1	Brown
2	Red
3	Orange
4	Yellow
5	Green
6	Blue
7	Violet
8	Grey
9	White

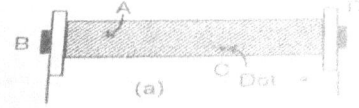

Fig.1

1. Fig.1(a), the body A, the end B and dot C are given in colours which represent the resistance. To find the value of such a resistance, suppose the Body is green, end is red and dot is orange. In such a resistance, the body colour gives the first significant figure, the end colour the second significant figure and the dot colour the number of zero followed by it. Hence the value of above resistance [Fig.1(a)] comes out to be

Body	End	Dot	
Green	Red	Orange	
5	2	000	$= 52000\Omega$

2. The resistances provided may be of the type as shown in Fig. 1(b). For such resistances the value is determined in the following manner: The colour of the first ring A on the extreme left represents the first significiant figure, the colour of second ring B the second significant figure and the colour of the third ring C the number of zeros followed by it.

As an example, if the first ring A is brown, second black and the third green then the given resistance will be

A	B	C
Brown	Black	Green
1	0	00000

$$= 1000000 \ \Omega$$

$$= 1 \text{ mega ohm.}$$

The colour of the ring D in both cases represents the percentage accuracy. If it is silver in colour the accuracy is $\pm$ 10% if it is gold in colour, the accuracy is $\pm$ 5 %.

Symbols Applicable in the Circuits of Electricity and Electronics

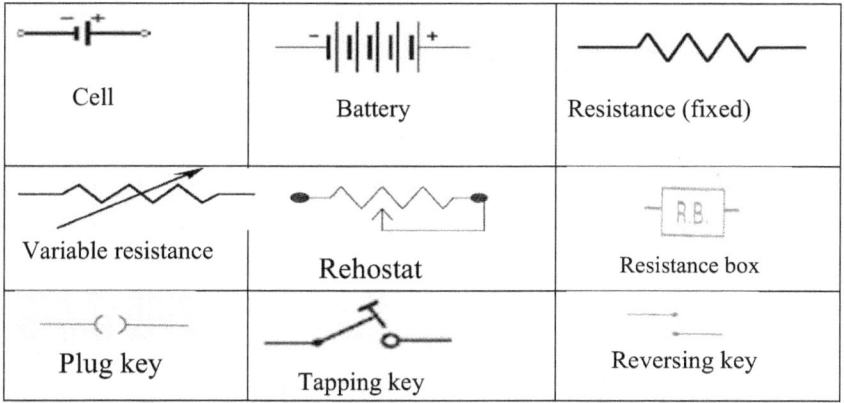

Cell	Battery	Resistance (fixed)
Variable resistance	Rehostat	Resistance box
Plug key	Tapping key	Reversing key

Morse key	Commutator	Galvanometer
Voltmeter	Ammeter	Shunted Galvanometer
Condenser	Electrolytic Condenser	Variable Condenser
Stepup Transformer	Stepdown Transformer	Earth connection
A.C. Signal	Coil (air core)	Coil (iron core)
Diode valve	Directly heated cathode	Indirectly heated cathode

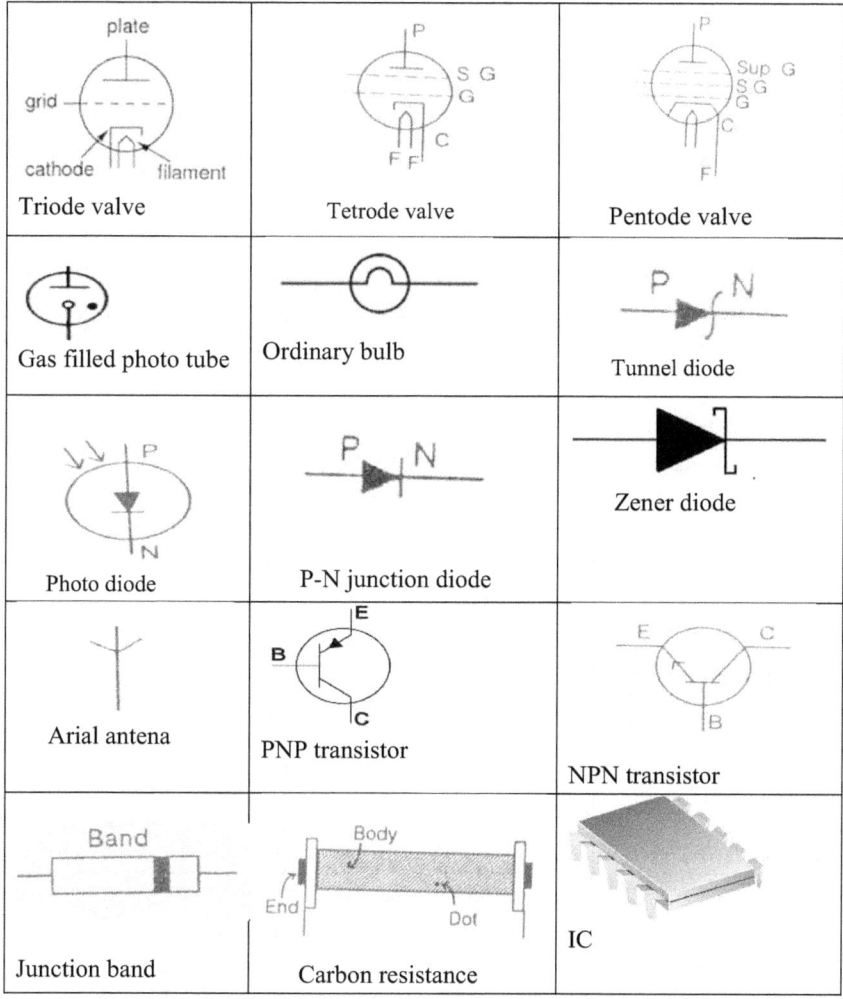

Triode valve	Tetrode valve	Pentode valve
Gas filled photo tube	Ordinary bulb	Tunnel diode
Photo diode	P-N junction diode	Zener diode
Arial antena	PNP transistor	NPN transistor
Junction band	Carbon resistance	IC

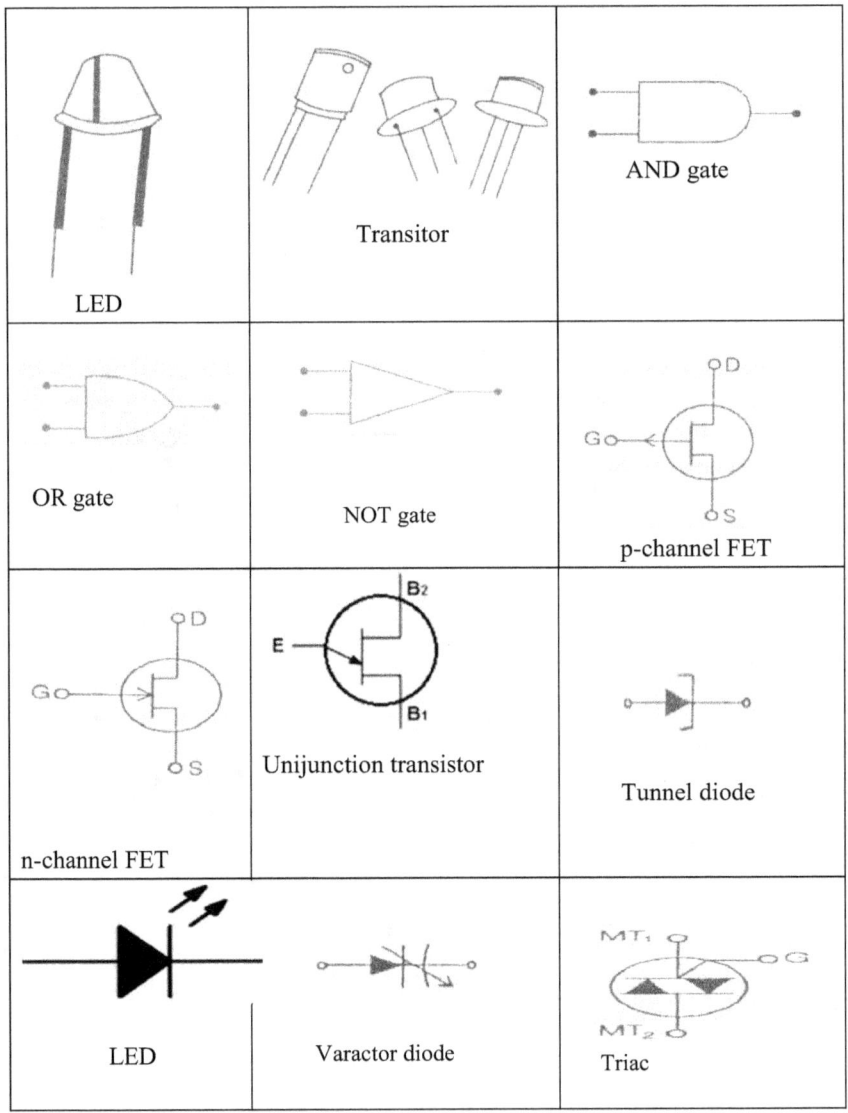

LED	Transitor	AND gate
OR gate	NOT gate	p-channel FET
n-channel FET	Unijunction transistor	Tunnel diode
LED	Varactor diode	Triac

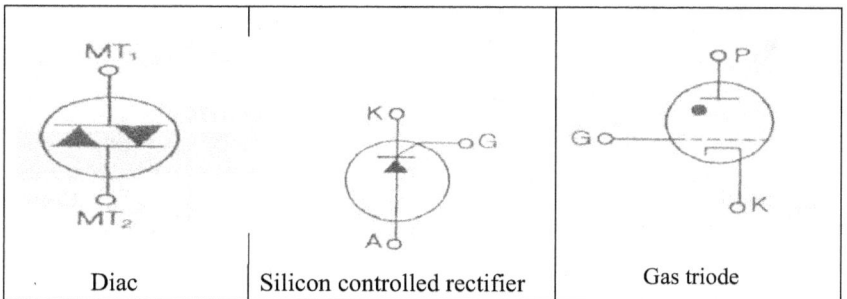

| Diac | Silicon controlled rectifier | Gas triode |